Nanogenerators and Self-Powered Systems

Edited by Bhaskar Dudem, Vivekananthan Venkateswaran and Arunkumar Chandrasekhar

Published in London, United Kingdom

Nanogenerators and Self-Powered Systems
http://dx.doi.org/10.5772/intechopen.100679
Edited by Bhaskar Dudem, Vivekananthan Venkateswaran and Arunkumar Chandrasekhar

Contributors
Ajay Kumar Singh, Musaab Zarog, Hossein Madadi Kojabadi, Liuchen Chang, Reza Ebrahimi, Sanusi Yekinni, Oyeshola Olayinka Hakeem, Ibrahim Asiata, Lawal Mubarak, Subramaniyam Narasimman, Lakshmi Narayanan Balakrishnan, Arunkumar Chandrasekhar, Zachariah C. Alex

First published in London, United Kingdom, 2023 by IntechOpen
IntechOpen is the global imprint of INTECHOPEN LIMITED, registered in England and Wales, registration number: 11086078, 5 Princes Gate Court, London, SW7 2QJ, United Kingdom

British Library Cataloguing-in-Publication Data
A catalogue record for this book is available from the British Library

Additional hard and PDF copies can be obtained from orders@intechopen.com

Nanogenerators and Self-Powered Systems
Edited by Bhaskar Dudem, Vivekananthan Venkateswaran and Arunkumar Chandrasekhar
p. cm.
Print ISBN 978-1-80356-254-4
Online ISBN 978-1-80356-255-1
eBook (PDF) ISBN 978-1-80356-256-8

Meet the editors

Bhaskar Dudem is currently working as a Post-Doctoral Research Fellow (from Oct 2019) at the Advanced Technology Institute (ATI), University of Surrey, England, UK. He received a Master of Technology (MTech., 2012) from the Department of Materials Science and Engineering, Indian Institute of Technology (IIT) Kanpur, India. Afterward, he was a teaching faculty member at K.L. University, India (2012–2014). In August 2018, he received a Ph.D. degree from the Department of Electronics and Radio Engineering, Institute for Wearable Convergence Electronics (IWCE), Kyung Hee University, South Korea. He also worked at the same institute as a Post-Doc for a year (Sep 2018–Sep 2019). His research interests include wearable and flexible piezo/triboelectric nanogenerators for energy harvesting and sensing applications, anti-reflection coatings, and solar, and hybrid energy cells, resulting in more than 30 works published in reputed peer-reviewed international journals.

Venkateswaran Vivekananthan is currently a Research Associate at Advanced Technology Institute (ATI), University Surrey. Prior to joining the University of Surrey, he worked as a Post-Doctoral Associate in the Faculty of Applied Energy Systems (Major in Mechatronics Engineering) at Jeju National University, South Korea. He was a recipient of the NRF Creative Challenge Support award as Principal-Investigator with project funding worth (USD 45,000/ year). He received his Ph.D. degree in February 2020 with the President Award for outstanding research and academics. His research focuses on energy harvesting using triboelectric and piezoelectric nanogenerators for self-powered sensor systems.

Arunkumar Chandrasekhar, Ph.D., is currently an Assistant Professor in the Department of Sensors and Biomedical Technology, Vellore Institute of Technology, India. He worked as a postdoctoral researcher at the Nanomaterials and Systems Laboratory, in South Korea. He obtained his Ph.D. in Mechatronics Engineering from Jeju National University, South Korea, where he was a recipient of a scholarship from the Korean Government Scholarship Program. Dr. Chandrasekhar also received the prestigious "Brain Korea 21+" Business Research Award from the Ministry of Education for excellence in research work. He is interested in wearable triboelectric nanogenerators, battery-free electronic devices, energy storage devices, microelectromechanical systems, and self-powered devices.

Contents

Preface

This Edited Volume is a collection of reviewed and relevant research chapters, concerning the developments within the "Nanogenerators and Self-Powered Systems" field of study. The book includes scholarly contributions by various authors and is edited by a group of experts pertinent to nanotechnology and nanomaterials. Each contribution comes as a separate chapter complete in itself but directly related to the book's topics and objectives.

The book includes chapters dealing with the topics:

1. Application of DC-DC Converters at Renewable Energy

2. A Study on Fiber Optic Temperature Sensor Using Al_2O_3 as High Index Overlay for Solar Cell Applications

3. Plant Base Renewable Energy to Power Nanoscale Sensors

4. Self-Powering Wireless Sensor Networks in the Oil and Gas Industry

5. Solar Photovoltaic Energy System

The target audience comprises scholars and specialists in the field.

InTechOpen

Chapter 1

Application of DC-DC Converters at Renewable Energy

Reza Ebrahimi, Hossein Madadi Kojabadi and Liuchen Chang

Abstract

Photovoltaics usually produce low voltage at their outputs. So, in order to inject their power into utility grids, the output voltage of solar panels should be increased to grid voltage level. Usually, the boost DC-DC converters will be connected between solar panels and grid-connected inverters to boost the panels' output voltage to more than 320 V (for 380/220 utilities). Various DC-DC converter topologies have been proposed in the past three decades to boost the photovoltaic panels' output voltage which will be discussed in this proposal. In order to increase the life span of photovoltaic panels, the DC-DC converts should absorb continuous low ripple current from solar panels. Maximum power point tracking (MPPT) is an algorithm implemented in photovoltaic (PV) inverters by DC-DC technology to continuously adjust the impedance seen by the solar array to keep the PV system operating at, or close to, the peak power point of the PV panel under varying conditions, like changing solar irradiance, temperature, and humidity. In this research work, various topologies of DC-DC converters that are suitable for renewable energy applications along with the advantages and disadvantages of control methods and the stability of converters with related control methods are discussed.

Keywords: DC-DC converters, photovoltaic cells, boost converters, nonisolated, MPPT, control methods

1. Introduction

Step-up DC-DC converter stores feed-in energy in magnetic field storage components like inductors, coupled inductors or electrical field storage components like capacitors and then flows it to the load with the higher voltage value compared to the feed-in voltage by using active and passive switching elements such as IGBTs, MOSFETs, and diodes. These converters have increasingly been used in many applications such as renewable energy sources and uninterruptable power supplies (UPS) [1–3].

A fundamental DC-DC converter is a simple PWM boost converter that is suitable for low-powerup to high-power and portable up to stationary devices. The major benefit of this converter that simplifies the modeling and implementation is lower element numbers. Higher efficiency, small size, lightweight, and reliable converters are strong demand for various applications. By increasing the duty cycle for reaching higher voltage gain, the voltage stress of active and passive elements increases, so the conversion efficiency is degraded. In practice, the voltage gain of conventional boost converters is limited due to the parasitic effects of MOSFETs or IGBTs and passive components [4–7].

Various voltage boost techniques such as charge pump, voltage multiplier, switched inductor, magnetic coupling, and multistage topologies were proposed for DC-DC converters.

Between charge pump topologies because of the modular structure, the popular topology is the switched capacitor topology. The critical issue of switched capacitor topology is the high-current transient that leads to lower efficiency and power density. To improve the current transient problem, an inductor has been added at the output of the converter to form a buck converter with the switches.

The advantages of switched capacitor technique are efficient regulation and elimination of current transients [8]. Hence, because of reduced power losses, the voltage and current spikes of converter are reduced, and the efficiency increases. Switched capacitor base converter can achieve high voltage gain with low voltage ratings of the output capacitance and low capacitance [9, 10].

Voltage multiplier topologies including a set of diodes and capacitors are of low-cost, efficient, and simple. Voltage multiplier converters divided into two types: (a) the voltage multiplier located in the middle of the circuit to reduce voltage stress, and (b) the voltage multiplier rectifier has been located at the output of the circuit to convert AC or pulse output voltage of the converter to DC voltage. The voltage stress of all parts of the circuit is lower than the output voltage which can be an advantage for this topology [11].

Also, high step-up DC-DC converters can be applied voltage lift and switched inductor technique. In this technique, capacitors will be charged to the input voltage and then the output voltage will be stepped up to the sum of voltage level of the capacitors. In these converters, the inductors are charged in parallel and discharged in series with the load [12].

Some new techniques proposed for high voltage gain DC-DC converters such as cascaded, multilevel, and interleaved in [13]. In quadratic boost, the first stage voltage stress is low and the switching frequency could be increased but the switching frequency limitation is for reducing the switching losses of the second stage [14]. So, the disadvantage of this converter is that the two stages of control are related to each other. Therefore, it can be concluded that this converter is suitable for low-power applications [15, 16]. Multilevel DC-DC converters because of high power and high voltage applications are suitable for industrial applications. The simplicity, flexibility, and modularity of a single-source multilevel DC-DC converter are the main advantages.

2. Nonisolated DC-DC converter

As expressed in the first section, DC-DC converters are widely utilized for renewable energy applications. Nonisolated DC-DC converter topologies in comparison with isolated converter topologies have a lot of advantages.

2.1 Buck topology

The basic DC-DC converter, the Buck converter, is shown in **Figure 1**. In this converter the power switch is connected between the input power supply and the load.

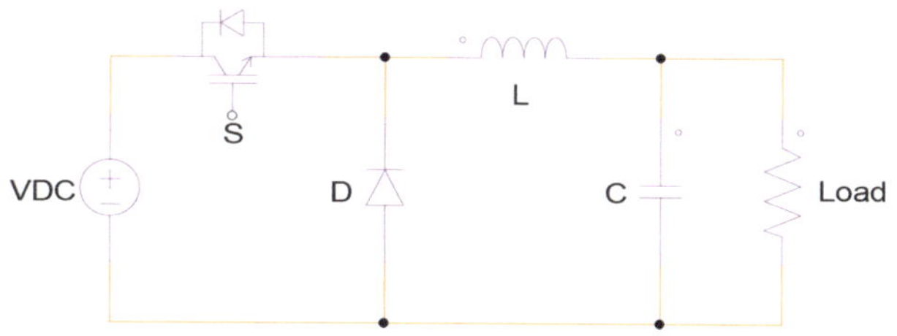

Figure 1.
Buck converter.

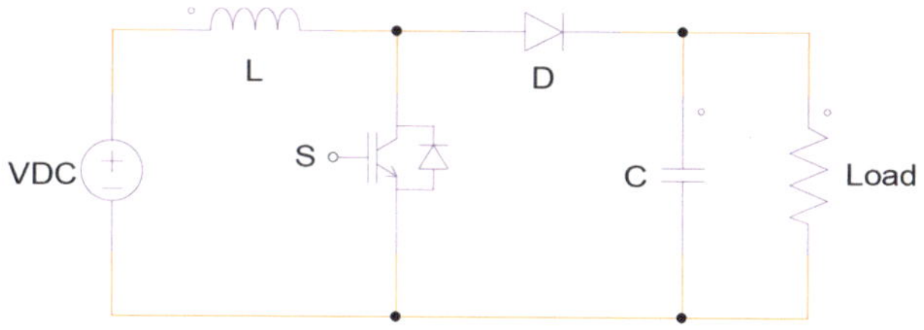

Figure 2.
Boost converter.

The power switch connects and disconnects periodically. The output voltage of the switch has a rectangular waveform that pass from a low pass filter and makes ripples DC output voltage. By assuming that the circuit elements are ideal, the conversion ratio of buck converter can be written as:

$$G_v = D \tag{1}$$

where D is the duty cycle of power switch. It is clear that the output voltage of converter can vary from zero to the input DC voltage.

2.2 Boost topology

The next topology is boost converter that is made from an interchanging inductor, switch, and diode, as shown in **Figure 2**. When the power switch is connected, the input current increases on the inductor. When the power switch is disconnected, the inductor current flow to the load through the diode. By assuming the ideal elements in circuit, the conversion ratio of boost converter can be expressed as:

$$G_v = {}^{1}\!/_{1-D} \tag{2}$$

where D is the duty cycle of power switch. It shows that the output voltage of converter can vary from input voltage up to higher voltages that limited by the parasitic elements of the circuit's active and passive components.

2.3 Buck-boost topology

Buck-boost converter is a mix of two different topologies as shown in **Figure 3**. The buck converter and the boost converter. The buck converter steps down and the

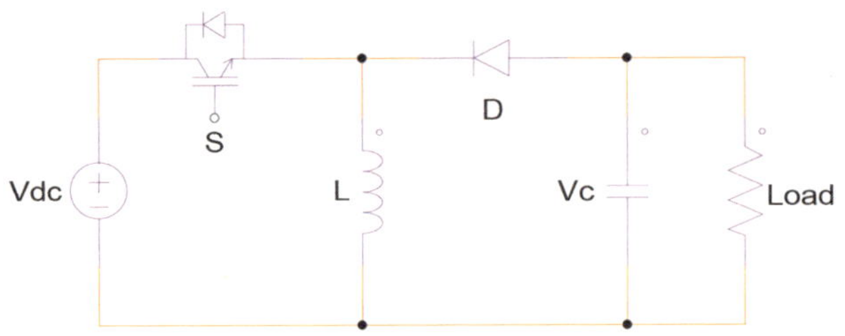

Figure 3.
Buck-Boost converter [17].

boost converter steps up the output voltage. This combined converter topology is used in many applications such as drive applications, stand-alone, and grid-connected photovoltaic (PV) systems. However, buck-boost converter is still under research to enhance the impression of the photovoltaic (PV) system. Researchers are working to increase the voltage gain of nonisolated DC-DC converters, as a result, many DC-DC converters are developed that include SEPIC, Cuk, Lou, and Z-source that all are based on buck-boost topology.

A novel topology of a double-switch buck-boost converter is proposed in [18]. It was shown experimentally that the converter is able to effectively track maximum power point for the photovoltaic application and also maintained maximum efficiency during load-varying conditions. Hybrid fuel cell-based system is using a coupled-inductor buck-boost converter. The proposed converter has higher efficiency, noninverting output, and low input and output ripples. Also, buck-boost converter is widely used in industrial applications. In [19], a bridgeless converter, i.e., the buck-boost converter for the motor drive application is proposed. In this topology, the converter and motor drive are integrated which leads to lower switching losses and conduction losses. Author in [20], proposed a boost-interleaved buck-boost converter, which consists of two switches, that is used for power factor correction application. This topology also decreases switching voltage stress, inductor losses and size of the magnetic interference. A cascade connection of two buck-boost converters that has one control switch has been utilized in the LED drive application. This topology leads to lower filter capacitance size [17]. In [21], a novel buck-boost converter for electrical vehicles proposed. The proposed converter controls the power transition between batteries and capacitors by using interleaved converter controlled by FPGA. In [22] a buck-boost converter is used to generate the telecommunication power system energy. In this converter, multiple input buck-boost converter is used that is between sources and DC-bus. This converter could reduce switching losses. In [23], a new technique was utilized for a smooth transition between switching modes of noninverter buck-boost converter.

2.4 SEPIC topology

Like buck-boost converter, single-ended primary inductor converter (SEPIC) is shown in **Figure 4**. SEPIC converter could step up and step down the output voltage. SEPIC topology could be utilized in various applications such as photovoltaic applications to improve the power factor and regulate the flickering DC voltage. Noninverting output of the SEPIC converter makes it more interesting than the buck-boost converter and it is suitable for high-power applications. To achieve high voltage output at SEPIC converter, duty cycle of the switch must be high.

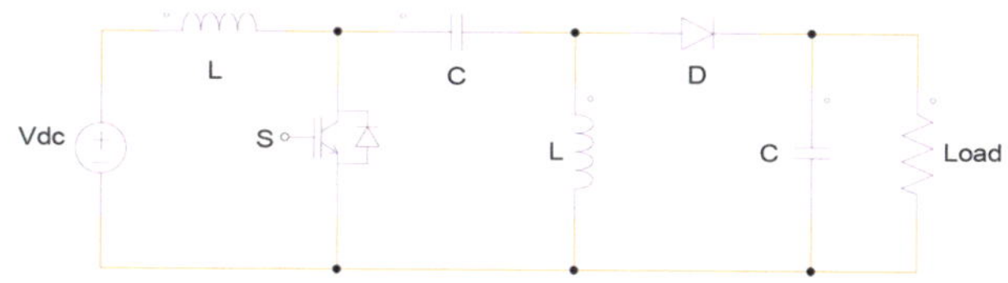

Figure 4.
SEPIC converter.

SEPIC topology is utilized in various applications. An inductor base SEPIC converter is used for renewable energy systems in [24]. The advantage of this topology include continuous input current and lower switching stress that leads to higher efficiency. In [25] a novel SEPIC converter is proposed that is suitable for high power factor correction. Proposed converter is an isolated bridgeless SEPIC and operates like a single-phase rectifier controlled with slide mode to achieve a high power factor. The advantage of this converter is low total harmonic distortion (THD) with suitable power factor correction. A modified converter of SEPIC proposed in [26] is a combination of two DC-DC converters and is suitable for renewable energy applications.

The advantage of this converter is low switching stress besides low input voltage and high output voltage. For photovoltaic application, an Integrated double-boost SEPIC converter is proposed in [27], which have a single switch and two inductors. This converter is able to provide high voltage gain with a low duty cycle. For increasing voltage gain and reducing voltage stress on the main switch a new topology of SEPIC was proposed in [28]. In this topology, SEPIC converter is combined with an inductor and two voltage multipliers. Because of continuous input current, it is suitable for renewable energy systems.

2.5 Cuk topology

As shown in **Figure 5**, Cuk converter topology is consist of a boost converter in the first stage and a buck converter in the second stage. The main application of Cuk converter is for voltage regulation and power factor correction (PFC) applications. Cuk converter is inverting converter that has an inverted output in compare to input voltage polarity. Also, this converter has lower switching losses that lead to higher efficiency. Voltage gain of the converter depends on duty cycle of switch. When the switch is ON, capacitors are discharged while inductors store energy. When the switch is OFF, the diode is conducted. In this topology, capacitors act as energy storage while in other converters inductors act as energy storage.

In [30], for the photovoltaic application, to achieve high voltage gain, the Cuk converter is coupled with switched inductor that leads to lower switching voltage.

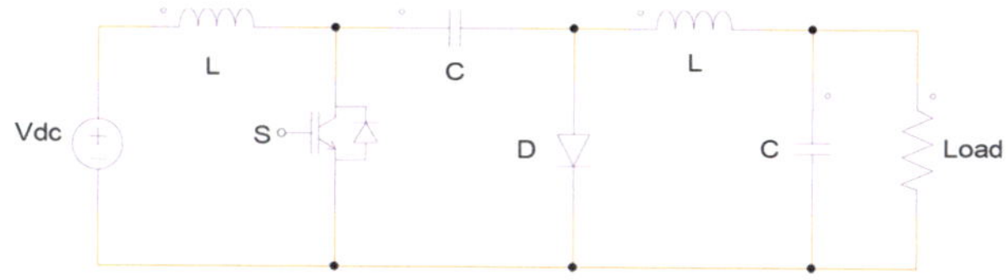

Figure 5.
Cuk converter [29].

In [31], a novel high step-up DC-DC Cuk converter that is used for fuel cell application is proposed that gives a wide range of duty cycle and high voltage gain. For improving the power factor correction and power quality of a motor drive of air-condition, a Cuk converter was proposed in [32]. Over a wide range of the operation, the converter was able to keep the power factor near one. In [33], a Cuk converter is proposed for power factor correction. The proposed bridgeless Cuk converter reduces the conduction losses and switching losses and could reduce the inductance size for power factor correction purposes.

In [34], a bridgeless Cuk converter for lighting application is proposed that has high efficiency with a low number of conductive components and low losses. In [35], a single-stage switched inductor Cuk converter for improving the efficiency of electrical bikes battery charging applications is proposed. This topology improves the efficiency, power factor, and total harmonic distortion of the converter.

2.6 Z-Source topology

One of the efficient topologies is the Z-Source topologies. Z-Source topology could step-up and step-down the voltage. As shown in **Figure 6**, its topology is based on inductor and capacitor network that connects the converter to the power source. Z-Source topology is mostly used in medium-power to high-power applications. Z-Source converter output ripple is low and the duty cycle is less than 0.5. At the same duty cycle, Z-Source converter have higher voltage boosting capability compared to conventional boost converter. Also, in compare to others, Z-Source converter has higher efficiency, and lower cost and size.

In [36], a new topology from combination of Z-source network, voltage multiplier, and flyback that achieve an efficiency of 89% has been proposed. This converter has higher component number compare to conventional Z-source topology. In [31], a modified Z-source converter proposed for photovoltaic application has been proposed. This converter has a common ground and the advantage include low switching stress and reduced size. In [37], a hybrid Z-source converter suitable for motor drive application was proposed. In [38], a Z-source converter controlled by sliding mode controller (SMC) was proposed for controlling a permanent magnet synchronous machine in electric traction application. Also, voltage adaption strategy was validated and the system efficiency improved. In [39], by using a z-source converter, the power factor correction of wireless power transfer applications for electrical vehicles and transportation improved. Power factor correction and regulation of the output voltage is done without using additional components only by applying the control circuit.

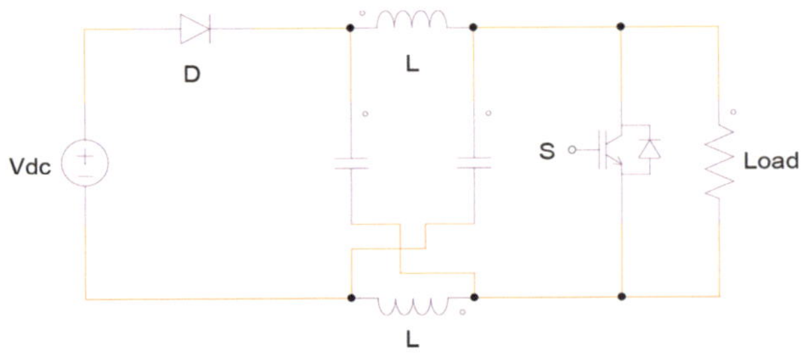

Figure 6.
Z-source converter [31].

2.7 Zeta topology

Zeta topology is well-known to be as a power optimizer. Zeta converters like SEPIC converter could be used in many applications like photovoltaic. Zeta converter in compare to other topologies, has a higher component count and higher complexity, asshown in **Figure** 7. It has the advantage of noninverting, regulated, and low ripple output voltage and continues current at output of converter. Some applications of Zeta converter are found, such as integration of Zeta converter with photovoltaic system to drive the BLDC water pump [40]. Also, this converter has fewer power losses and implements maximum power point tracking from PV cells. A constant output voltage under load-varying condition of wind turbine application is achieved by Zeta converter in [41]. A novel modified Zeta topology is used for high voltage conversion that improve efficiency in high voltage applications proposed in [42, 43]. In [44], a smart combination of Zeta converter and SEPIC converter is proposed for plugin electric vehicles that could operate in three modes, i.e., regenerative, propulsion, and plugin charging modes. This converter prepares the capability of voltage gain in all modes which leads to increasing the efficiency of electrical vehicles.

2.8 Recently developed nonisolated topologies

Nowadays, nonisolated topologies are used abundantly in many modern applications. Traditional nonisolated converters have lower efficiency and reduce the life span of electrical parts and systems in comparison to recently proposed topologies. So, the combination of topologies is an active manner to propose new topologies. Combination of topologies is just based on the advantages and disadvantages of topologies. Important parameters of nonisolated topologies are input and output ripples, continuous and discontinuous input and output current, switching voltage and current stress and duty cycle of switches. Also, **Table 1** shows the advantages and disadvantages of these converters. **Figure 6**, shows recently proposed common ground nonisolated DC-DC topologies. **Figure 8(a)** presents a DC-DC topology that could step up and step down the output voltage [58]. It is suitable for photovoltaic applications. For increasing the voltage gain, it uses dual coupled inductors in series. Also, it works on low-duty cycle for preparing high voltage gain.

Figure 8(b) proposed a high-gain DC-DC topology with parallel input and series output DC-DC topology. It has dual coupled inductors with a voltage multiplier [59]. The output of the converter is composed of interleaved series-connected capacitors while the input side is composed of two inductors that connect in parallel to share the input current and voltage ripple. This topology is used in both industrial and domestic applications. Proposed converter has normal switching stress, lower output voltage ripple and high voltage gain. **Figure 8(c)** shows a transformer-less high-gain DC-DC

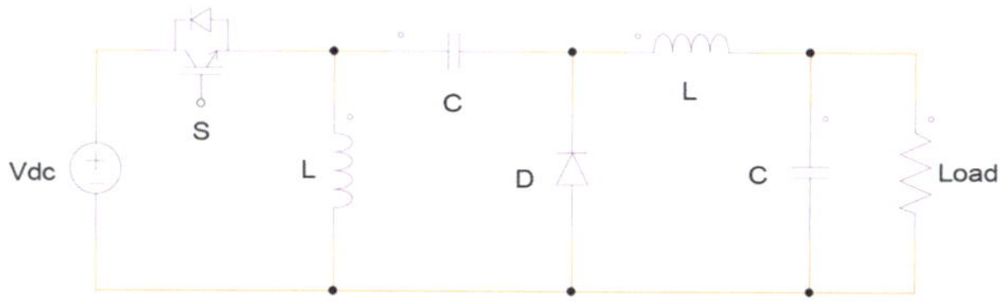

Figure 7.
Zeta converter.

Topology	Features	Benefits	Limitations
Buck Boost [18, 29, 45–47]	• Low complexity • Small size • Simple Control • Low Cost	• Suitable for low-power application • High switching frequency	• High output ripple • Discontinues output current
SEPIC [24, 25, 48–50]	• Low complexity • Small size • Simple Control • Low Cost	• Noninverting Output voltage • utilized as PFC	• Low voltage gain • Complex control in multiinput and multioutput
CUK [30, 31, 36, 51–53]	• Low complexity • Small size • Simple Control • Low Cost	• Suitable for low-power application	• Inverted Output voltage • Discontinues output current
Z-Source [31, 36, 37, 54, 55]	• Medium complexity • Small size • Simple Control • Low Cost	• Noninverting Output voltage	• Unidirectional power flow • Discontinues input current
Zeta [40, 45, 56, 57]	• Medium complexity • Small size • Simple Control • Low Cost	• Suitable for medium and high-power application • Noninverting Output voltage	• Unidirectional power flow
[58]	• Medium complexity • Small size • Complex Control • High Cost	• Noninverting Output voltage • Common ground • Renewable energy application	• Input conduction losses (Coupled Inductor)
[59]	• Medium complexity • Medium size • Complex Control • High Cost	• Noninverting Output voltage • Renewable energy application	• Unidirectional power flow
[60]	• Medium complexity • Medium size • Complex Control • High Cost	• Noninverting Output voltage • Renewable energy application	• Unidirectional power flow

Table 1.
DC-DC converter topologies compare.

topology, as proposed in [60]. The advantage of this converter is low component counts and high efficiency. This converter uses the component best which leads to higher voltage gain with lower component counts. So, it is not necessary to use circuits such as: voltage lift, voltage multiplier, and coupled transformers. This topology is used for DC-microgrid, and renewable energy systems like photovoltaic cells.

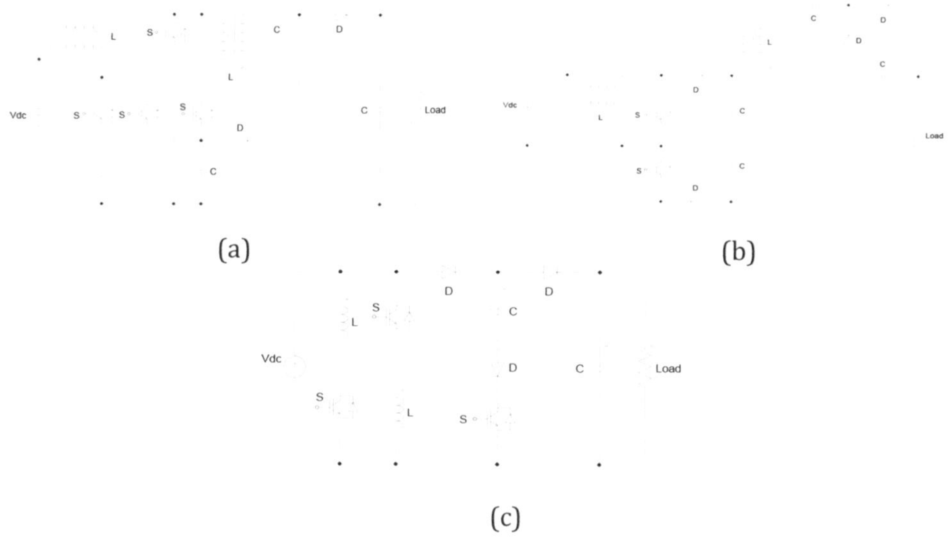

Figure 8.
Novel nonisolated topologies (a) [58], (b) [59], (c) [60].

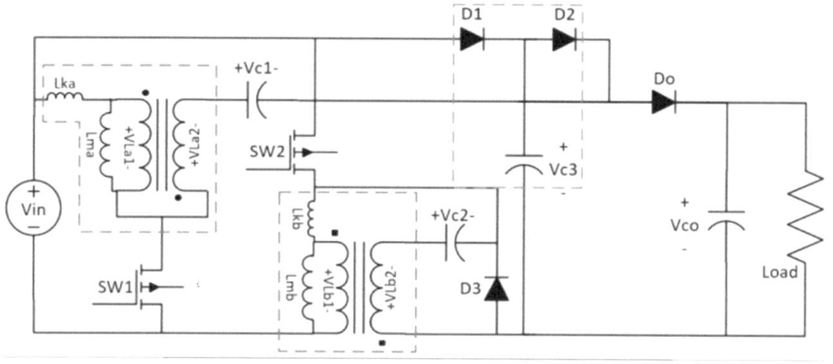

Figure 9.
Proposed topology of [1].

In [1], a novel step-up coupled inductor-based DC-DC converter has been proposed. The topology consists of two coupled inductors, and two power switches that work simultaneously. **Figure 9** shows the topology of the proposed converter. The conversion ratio of the converter can be expressed as:

$$G_V = \frac{V_o}{V_{in}} = \frac{3 + 2N_b + N_a}{1 - D} \tag{3}$$

where $N_a = {}^{V_{La2}}/_{V_{La1}}, N_b = {}^{V_{Lb2}}/_{V_{Lb1}}$ and D is the duty cycle.

The proposed topology is simulated in PSIM software and the related results are shown in **Figure 10**. Real-time (experimental) results of the related topology with the parameters that are written in **Table 2** are shown in **Figure 11**. for the input voltage of 11 V. So, the voltage gain is 268/11 = 24.36.

Switching frequency	$f_s = 50\text{kHz}$
Coupled inductors	
$L_a = 196\mu H$	$L_b = 196\mu H$
$N_a = 1$	$N_b = 1$
Capacitors	
$C_1 = C_3 = 220\mu F/100V$ $C_2 = 220\mu F/50V$	$C_o = 1500\mu F/400V$
Switches & Fast diodes	
$S_1 = S_2 = IRFP150$ $(100V, 29A, R_{DS(on)} = 0.055\ \Omega)$	$D_1 = D_3 =$ FR303 (200V, 3A) $D_2 = D_o =$ MUR840 (400V, 8A)

Table 2.
Parameters of the experimental setup of [1].

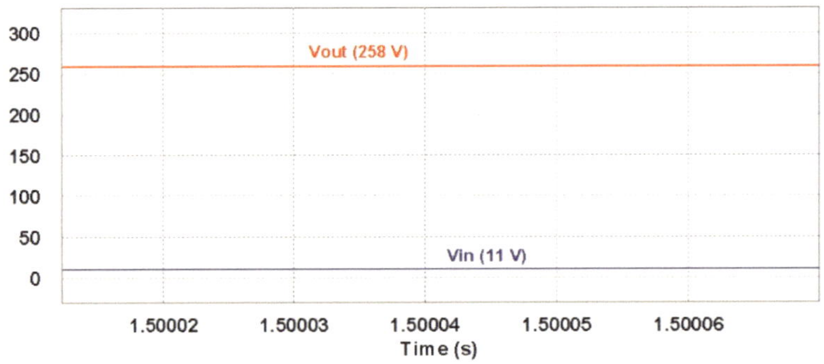

Figure 10.
Simulated input and output voltage of proposed topology in [1].

3. Control techniques

Control techniques are essential for achieving maximum efficiency in DC-DC topologies because these techniques could optimize the operation of the converters. The parameters that we could control include: input and output voltage, duty cycle, and reference signal. For achieving the high voltage gain, the controller increases the duty cycle to step up the voltage and for achieving the lower voltage gain, the controller decreases the duty cycle to step down the voltage by considering the reference signal. All scenarios of the controller are applied to achieve optimum control on DC-DC converters to get the required output [61]. Features like response time, efficiency could be controlled by some control techniques [62, 63] but all features cannot be achieved simultaneously. Designer must trade off due to the special application.

3.1 PID control

The first and the most common control technique that has been applied in industry is proportional integral derivate (PID) control which is accepted for various applications like motor drive and renewable energy systems. PID controller is preferred due

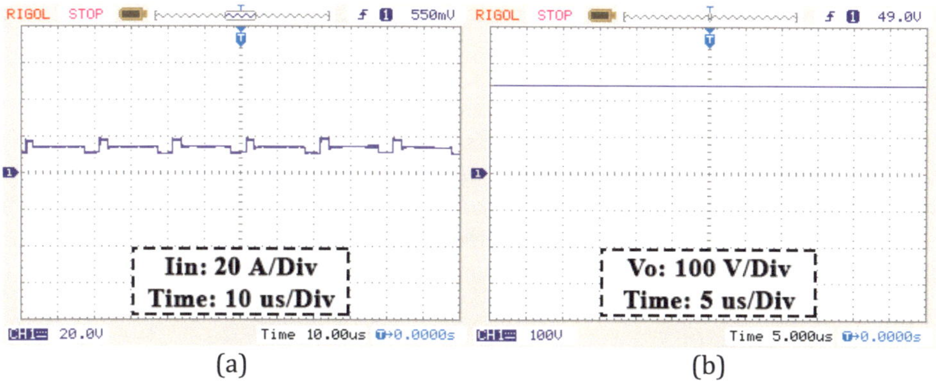

Figure 11.
Real-time results (a) input current, and (b) output voltage of proposed topology in [1] for input voltage of 11 V.

to simple implementation and the robust response over a wide range of operating, as shown in **Table 3**. For controlling the DC-DC converter, PID controller is conventional and effective technique that take the feedback signal from the output and control the duty cycle of the switch to achieve the required voltage gain. Its efficiency is constant in various applications. The main advantages of PID controllers are easy implementation and low complexity. Novel hybrid control techniques are proposed in [77] that improve the control and efficiency of system in renewable energy applications.

3.2 Sliding mode control

Sliding mode controller (SMC) is a nonlinear discontinuous controller. SMC controller is suitable for applications that have external disturbances. To reduce the error

Controller	Features	Benefits	Limitations
PID [47, 52, 64, 65]	• Easy to Implement & Low complexity	• Fast transient response • Easy to combine with other control strategies	• High steady-state error and overshoot
SMC [52, 66–68]	• Robust & Nonlinear Control	• Fast Dynamic • Fast settling time • Robust	• Chattering Problem • Large Overshoot
MPC [47, 69–71]	• Robust & Nonlinear Control • Predict future state	• Fast Response • Efficient tracking	• Depends on parameters • High calculation
SSM [72, 73]	• Robust & Nonlinear Control • Suitable for MIMO systems	• Better transient response • Lower Overshoot by load varying	• Large initial time • Needs model details
FLC [73–76]	• Robust & Nonlinear Control • High Stability	• Low Overshoot • Efficient tracking response • Without mathematical model	• High calculation • Needs rules to operate • Large settling time

Table 3.
Control techniques compare.

between desired and output voltages, the system converges to the sliding surface. For continued stability, the system tends to slide on a sliding surface. The error found in this operation calls the chattering effect [66, 78]. Before the system tends to a constant sliding mode, the oscillation calls chattering effect. Mixing other control techniques with SMC leads to overcoming this problem. The principle of this control method is to reach the output signal equal or close to the reference value on the sliding surface. **Figure 12** shows the SMC control system on a DC-DC converter. The feedback signal is read by the SMD controller to produce the required signal and generate the switching signals.

3.3 Model predictive control

Novel control method that is not conventional is model predictive control (MPC). The principle of this controller is getting feedback to control algorithm. This controller method uses model of the system to predict the next state value to create suitable signals for controlling the converter [69]. MPC controller is capable to control multiinput, multioutput systems (MIMO). As an advantage, the MPC controller is a multivariable controller. MPC could manage the outputs and inside system parameters. Also, the MPC controller could combine with the minimization of cost function, operating cost, economic load dispatch and optimized power flow management.

Due to the intermittent changes occurring and uncertain behavior of the DC-DC converter, controlling the converter is a critical task. Implementation of model predictive control on DC-DC converter shown in **Figure 13**. Feedback from output signal of DC-DC converter with utilization of predictive control algorithm, predict the next state of the converter and send the related signal to operate the converter smoothly. Turbulences could occur in the input or output side of the converter.

3.4 State space modeling

The mathematical modeling of a real system by means of inputs, outputs, state variables, and equations is state space modeling (SSM). SSM of a real system is represented by state equations which are two types of equations that call state equations. The number of equations and the order of state space model depends on the number of inputs/outputs that include the physical system [80, 81]. The state space model is able to represent higher-order real systems in the time domain simply.

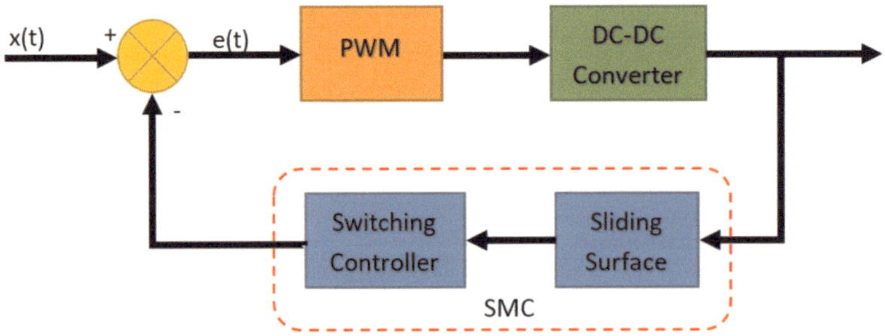

Figure 12.
SMC integration with DC-DC converter [25].

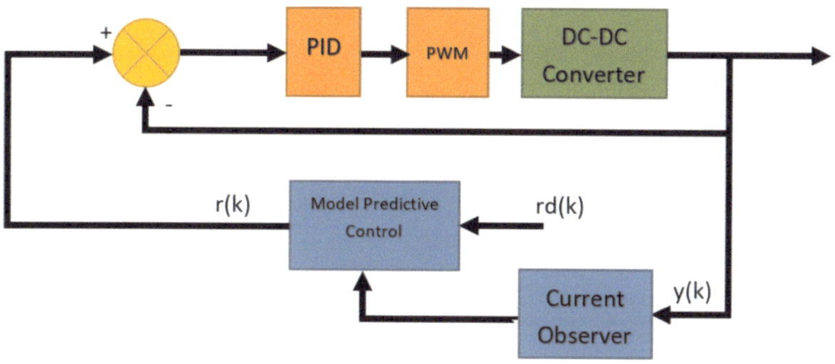

Figure 13.
Controlling of the DC-DC converter using the MPC control technique [79].

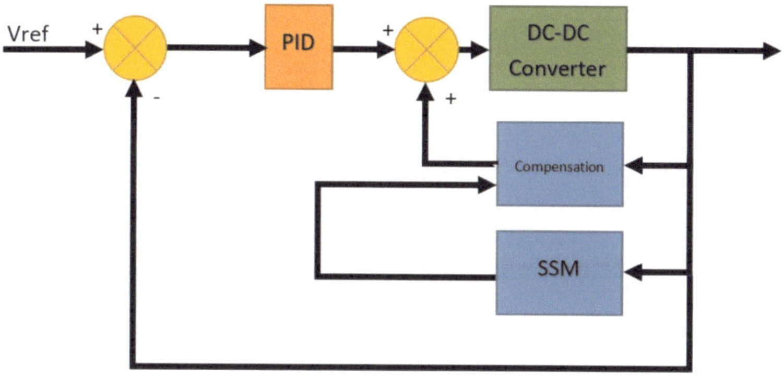

Figure 14.
SSM control for DC-DC converter.

Since just fundamental representations of real system are necessary for state space modeling, it can be used for nonlinear and multiinput, multioutput systems [51].

The operation of the state space modeling technique controller on DC-DC converter is shown in **Figure 14**. SSM is a mathematical controller that uses a mathematical model to control the system in different states that have higher efficiency compared to other methods. The advantage of this method is in reducing the order of complex system which leads to minimizing the computational time of the controller. Feedback control loop of state space modeling implemented on DC-DC converter is shown in **Figure 14**. In high-precision needed systems, SSM control is suitable.

3.5 Fuzzy logic control

The newest controller in category of nonconventional and nonlinear controllers is the fuzzy logic controller (FLC). FLC works like human thought process. Predefined rules are required to implement the human thought process. Membership function is lingual rules that define the input and output of the system. FLC does not need any mathematical model, so, it is much simpler than SSM and MPC. Also, the FLC controller could implement on nonlinear systems. FLC controller takes the feedback of system's crisp value, change it to lingual form and compares it with the membership functions which calls fuzzification. After the process, convert the lingual phrases back to the crisp value that is named defuzzification. For nonlinear control systems that

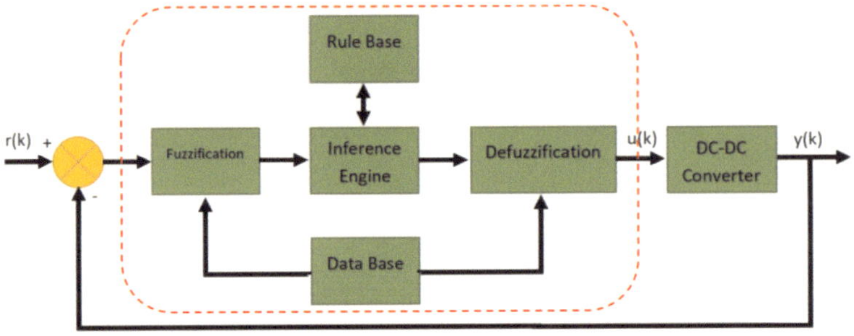

Figure 15.
Fuzzy logic control for DC-DC Converter [76].

have vague boundary conditions, FLC has an efficient response. The disadvantage of this control method is higher computational time. To resolve this problem, FLC is combined with other control techniques and works in offline mode [74].

Because of the higher effectivity of the FLC controller, it could be used in domestic and industrial applications. In [82, 83], an FLC controller applied to automatic car brake system charge controller in electrical vehicles. Also, in [84] it is used for controlling of marine surface vessels and underwater vehicles. On the other hand, FLC is used for industrial applications and the power generation systems [75, 85, 86]. In a grid-connected inverter, the FLC controller is used for multiinput DC-DC converter to operate it in boost mode [87]. In [88], FLC implemented on integration of photovoltaic panels (PV) with SEPIC topology to increase efficiency. The control algorithm of a DC-DC converter by FLC is shown in **Figure 15**. The feedback of controller can be obtained from the output of DC-DC converter.

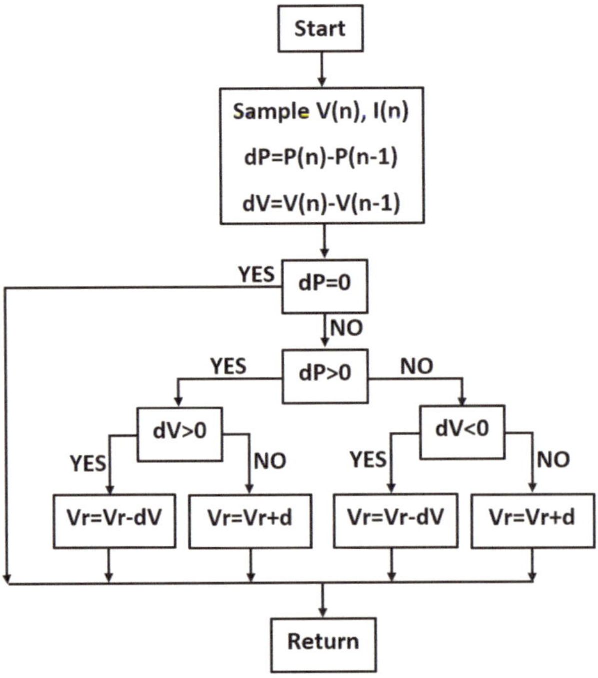

Figure 16.
Algorithm of P&O.

3.6 MPPT Algorithm

To improve the efficiency of photovoltaic energy systems, PV modules must operate at maximum power points to deliver the maximum power to the load. Some MPPT techniques are used to deliver the maximum energy of solar to the load and batteries. Different techniques of maximum power point tracking methods have been proposed like open circuit voltage method [89, 90], short circuit current method [91, 92], fuzzy logic method [93, 94], perturb and observe method [95–97], and incremental conductance method [98–100]. The most popular method is perturbed and observes (P&O) method that its algorithm is shown in **Figure 16**. In this method, a small perturb on duty cycle is applied to cause power variation and the output power of PV is measured. If the new measured power is higher than the last measured power the perturbation is continued in this direction otherwise the perturbation is continuing in the reverse direction. In this algorithm, when voltage increase leads to an increase in the power, it means that the operating point of modules is on the left side of the peak of MPPT diagram and when voltage increase leads to decrease in the power, it means that the operating point of modules is on the right side of the peak of MPPT diagram.

4. Conclusions

The purpose of this research work is based on the performance study of DC-DC converter topologies that are applicable in renewable energy systems. Advantages and disadvantages of converters are discussed with their applications. Also, the advantage and disadvantages of control methods and stability of converters with related control methods in renewable energy application are explained. Both conventional and novel DC-DC converters are discussed in this chapter with their advantages and disadvantages. The buck-boost, Cuk, SEPIC, Z-Source, and Zeta topologies in conventional category and some new proposed topologies in novel category are explained. The converters are compared in terms of voltage gain, voltage stress over switches and diodes and the number of components.

Author details

Reza Ebrahimi[1]*, Hossein Madadi Kojabadi[1] and Liuchen Chang[2]

1 Sahand University of Technology, Tabriz, Iran

2 University of New Brunswick, Fredericton, Canada

*Address all correspondence to: r_ebrahimi@sut.ac.ir

References

[1] Ebrahimi R, Madadi Kojabadi H, Chang L, Blaabjerg F. Coupled-inductor-based high step-up DC–DC converter. IET Power Electronics. 2019;**12**(12): 3093-3104. DOI: 10.1049/iet-pel.2018.6151

[2] Ghaffarpour M, Ebrahimi R, Kojabadi HM, Chang L, Guerrero JM. Novel high voltage gain dc–dc converter with dynamic analysis. IET Power Electronics. 2021;**14**(3):562-583. DOI: 10.1049/pel2.12028

[3] Kojabadi HM, Ebrahimi R, Esmaeilifard H, Chang L, Chen Z, Blaabjerg F. High boost transformer-based Z-source inverter under continuous input current profile. IET Power Electronics. 2019;**12**(14): 3716-3723. DOI: 10.1049/iet-pel.2019.0366

[4] Hoseinzadeh Lish M, Ebrahimi R, Madadi Kojabadi H, Guerrero JM, Nourani Esfetanaj N, Chang L. Novel high gain DC–DC converter based on coupled inductor and diode capacitor techniques with leakage inductance effects. IET Power Electronics. 2020; **13**(11):2380-2389. DOI: 10.1049/iet-pel.2020.0117

[5] Hoseinzadeh M, Ebrahimi R, Kojabadi HM. A cascade high gain DC-DC converter employing coupled inductor and diode capacitor. In: 2019 5th Conference on Knowledge Based Engineering and Innovation (KBEI). Tehran, Iran: IEEE; 2019. pp. 205-209. DOI: 10.1109/KBEI.2019.8735079

[6] Ma W, Xue X, Liu G. Techno-economic evaluation for hybrid renewable energy system: Application and merits. Energy. 2018; **159**:385-409. DOI: 10.1016/j.energy.2018.06.101

[7] Acar C, Dincer I. Environmental impact assessment of renewables and conventional fuels for different end use purposes. International Journal of Global Warming. 2017;**13**(3–4):260-277. DOI: 10.1504/IJGW.2017.087197

[8] Singh R, Bansal RC. Review of HRESs based on storage options, system architecture and optimisation criteria and methodologies. IET Renewable Power Generation. 2018;**12**(7):747-760. DOI: 10.1049/iet-rpg.2017.0603

[9] Al-Ammar EA et al. Residential community load management based on optimal design of standalone HRES with model predictive control. IEEE Access. 2020;**8**:12542-12572. DOI: 10.1109/ACCESS.2020.2965250

[10] Vieira MDM, Huijbregts MAJ. Comparing mineral and fossil surplus costs of renewable and non-renewable electricity production. International Journal of Life Cycle Assessment. 2018; **23**(4):840-850. DOI: 10.1007/s11367-017-1335-6

[11] Day C, Day G. Climate change, fossil fuel prices and depletion: The rationale for a falling export tax. Economic Modelling. 2017;**63**:153-160. DOI: 10.1016/j.econmod.2017.01.006

[12] Burmester D, Rayudu R, Seah W, Akinyele D. A review of nanogrid topologies and technologies. Renewable and Sustainable Energy Reviews. 2017; **67**:760-775. DOI: 10.1016/j.rser.2016.09.073

[13] Soliman MA, Hasanien HM, Alkuhayli A. Marine predators algorithm for parameters identification of triple-diode photovoltaic models. IEEE Access. 2020;**8**:155832-155842. DOI: 10.1109/ACCESS.2020.3019244

[14] Colmenar-Santos A, Monteagudo-Mencucci M, Rosales-Asensio E, de Simón-Martín M, Pérez-Molina C. Optimized design method for storage systems in photovoltaic plants with delivery limitation. Solar Energy. 2019; **180**:468-488. DOI: 10.1016/j.solener. 2019.01.046

[15] Sabouhi H, Abbaspour A, Fotuhi-Firuzabad M, Dehghanian P. Reliability modeling and availability analysis of combined cycle power plants. International Journal of Electrical Power & Energy Systems. 2016;**79**:108-119. DOI: 10.1016/j.ijepes.2016.01.007

[16] Hernández-Callejo L, Gallardo-Saavedra S, Alonso-Gómez V. A review of photovoltaic systems: Design, operation and maintenance. Solar Energy. 2019;**188**:426-440. DOI: 10.1016/j.solener.2019.06.017

[17] Alonso JM, Vina J, Vaquero DG, Martinez G, Osorio R. Analysis and design of the integrated double buck–boost converter as a high-power-factor driver for power-LED lamps. IEEE Transactions on Industrial Electronics. 2012;**59**(4):1689-1697. DOI: 10.1109/ TIE.2011.2109342

[18] Kaouane M, Boukhelifa A, Cheriti A. Regulated output voltage double switch Buck-Boost converter for photovoltaic energy application. International Journal of Hydrogen Energy. 2016;**41**(45): 20847-20857. DOI: 10.1016/j.ijhydene. 2016.06.140

[19] Bist V, Singh B. An adjustable-speed pfc bridgeless buck–boost converter-Fed BLDC motor drive. IEEE Transactions on Industrial Electronics. 2014;**61**(6): 2665-2677. DOI: 10.1109/TIE.2013. 2274424

[20] Chen J, Maksimovic D, Erickson RW. Analysis and design of a low-stress buck-boost converter in universal-input PFC applications. IEEE Transactions on Power Electronics. 2006;**21**(2):320-329. DOI: 10.1109/ TPEL.2005.869744

[21] Blanes JM, Gutiérrez R, Garrigós A, Lizán JL, Cuadrado JM. Electric vehicle battery life extension using ultracapacitors and an FPGA controlled interleaved buck–boost converter. IEEE Transactions on Power Electronics. 2013; **28**(12):5940-5948. DOI: 10.1109/ TPEL.2013.2255316

[22] Onar OC, Shirazi OHA, Khaligh A. Grid interaction operation of a telecommunications power system with a novel topology for multiple-input buck-boost converter. IEEE Transactions on Power Delivery. 2010;**25**(4): 2633-2645. DOI: 10.1109/TPWRD.2009. 2031490

[23] Lee Y-J, Khaligh A, Emadi A. A compensation technique for smooth transitions in a noninverting buck–boost converter. IEEE Transactions on Power Electronics. 2009;**24**(4):1002-1015. DOI: 10.1109/TPEL.2008.2010044

[24] Moradpour R, Ardi H, Tavakoli A. Design and implementation of a new SEPIC-based high step-Up DC/DC converter for renewable energy applications. IEEE Transactions on Industrial Electronics. 2018;**65**(2): 1290-1297. DOI: 10.1109/TIE.2017. 2733421

[25] Lopez-Santos O, Cabeza-Cabeza AJ, Garcia G, Martinez-Salamero L. Sliding mode control of the isolated bridgeless SEPIC high power factor rectifier interfacing an AC source with a LVDC distribution bus. Energies. 2019;**12**(18). DOI: 10.3390/en12183463

[26] Gules R, dos Santos WM, dos Reis FA, Romaneli EFR, Badin AA.

A modified SEPIC converter with high static gain for renewable applications. IEEE Transactions on Power Electronics. 2014;**29**(11):5860-5871. DOI: 10.1109/TPEL.2013.2296053

[27] Sabzali AJ, Ismail EH, Behbehani HM. High voltage step-up integrated double Boost–Sepic DC–DC converter for fuel-cell and photovoltaic applications. Renewable Energy. 2015; **82**:44-53. DOI: 10.1016/j.renene.2014.08.034

[28] Ardi H, Ajami A. Study on a high voltage gain SEPIC-based DC–DC converter with continuous input current for sustainable energy applications. IEEE Transactions on Power Electronics. 2018; **33**(12):10403-10409. DOI: 10.1109/TPEL.2018.2811123

[29] Mumtaz F, Zaihar Yahaya N, Tanzim Meraj S, Singh B, Kannan R, Ibrahim O. Review on non-isolated DC-DC converters and their control techniques for renewable energy applications. Ain Shams Engineering Journal. 2021;**12**(4):3747-3763. DOI: 10.1016/j.asej.2021.03.022

[30] de Morais JCDS, de Morais JLDS, Gules R. Photovoltaic AC module based on a cuk converter with a switched-inductor structure. IEEE Transactions on Industrial Electronics. 2019;**66**(5): 3881-3890. DOI: 10.1109/TIE.2018.2856202

[31] Babaei E, Abu-Rub H, Suryawanshi HM. Z-Source converters: Topologies, modulation techniques, and application–Part I. IEEE Transactions on Industrial Electronics. 2018;**65**(6): 5092-5095. DOI: 10.1109/TIE.2018.2793738

[32] Singh B, Bist V. Improved power quality bridgeless Cuk converter fed brushless DC motor drive for air conditioning system. IET Power Electronics. 2013;**6**(5):902-913. DOI: 10.1049/iet-pel.2013.0050

[33] Fardoun AA, Ismail EH, Sabzali AJ, Al-Saffar MA. New efficient bridgeless cuk rectifiers for PFC applications. IEEE Transactions on Power Electronics. 2012; **27**(7):3292-3301. DOI: 10.1109/TPEL.2011.2182662

[34] Saravanan D, Gopinath M. A novel power factor correction modified bridge less-CUK converter for LED lamp applications. International Journal of Power Electronics and Drive Systems. 2016;**7**:880. DOI: 10.11591/ijpeds.v7.i3.pp880-891

[35] Ananthapadmanabha BR, Maurya R, Arya SR. Improved power quality switched inductor cuk converter for battery charging applications. IEEE Transactions on Power Electronics. 2018; **33**(11):9412-9423. DOI: 10.1109/TPEL.2018.2797005

[36] Torkan A, Ehsani M. A novel nonisolated Z-source DC–DC converter for photovoltaic applications. IEEE Transactions on Industry Applications. 2018;**54**(5):4574-4583. DOI: 10.1109/TIA.2018.2833821

[37] Ellabban O, Abu-Rub H. An overview for the Z-Source converter in motor drive applications. Renewable and Sustainable Energy Reviews. 2016; **61**:537-555. DOI: 10.1016/j.rser.2016.04.004

[38] Battiston A, Miliani E-H, Martin J-P, Nahid-Mobarakeh B, Pierfederici S, Meibody-Tabar F. A control strategy for electric traction systems using a PM-motor fed by a bidirectional Z-source inverter. IEEE Transactions on Vehicular Technology. 2014;**63**(9):4178-4191. DOI: 10.1109/TVT.2014.2312434

[39] González-Santini NS, Zeng H, Yu Y, Peng FZ. Z-Source resonant converter with power factor correction for wireless power transfer applications. IEEE Transactions on Power Electronics. 2016; **31**(11):7691-7700. DOI: 10.1109/TPEL.2016.2560174

[40] Kumar R, Singh B. BLDC motor-driven solar PV array-fed water pumping system employing zeta converter. IEEE Transactions on Industry Applications. 2016;**52**(3): 2315-2322. DOI: 10.1109/TIA.2016.2522943

[41] Sowmya B, Saranya D. Solar integrated ZETA converter for DFIG based wind energy conversion system applications. AIP Conference Proceedings. 2022;**2405**(1):40003. DOI: 10.1063/5.0073305

[42] Andrade AMSS, Martins MLDS. Quadratic-boost with stacked zeta converter for high voltage gain applications. IEEE Journal of Emerging and Selected Topics in Power Electronics. 2017;**5**(4):1787-1796. DOI: 10.1109/JESTPE.2017.2706220

[43] Hosseini SH, Alishah RS, Kurdkandi NV. Design of a new extended zeta converter with high voltage gain for photovoltaic applications. In: 2015 9th International Conference on Power Electronics and ECCE. Asia (ICPE-ECCE Asia). IEEE; 2015. pp. 970-977. DOI: 10.1109/ICPE.2015.7167899

[44] Pathak MK. Single-stage ZETA-SEPIC-based multifunctional integrated converter for plug-in electric vehicles. IET Electrical Systems in Transportation. 2018;**8**(2):101-111. DOI: 10.1049/iet-est.2017.0063

[45] Singh SN. Selection of non-isolated DC-DC converters for solar photovoltaic system. Renewable and Sustainable Energy Reviews. 2017;**76**:1230-1247. DOI: 10.1016/j.rser.2017.03.130

[46] Ramírez-Murillo H et al. An efficiency comparison of fuel-cell hybrid systems based on the versatile buck–boost converter. IEEE Transactions on Power Electronics. 2018;**33**(2): 1237-1246. DOI: 10.1109/TPEL.2017.2678160

[47] Xu Q, Vafamand N, Chen L, Dragičević T, Xie L, Blaabjerg F. Review on advanced control technologies for bidirectional DC/DC converters in DC microgrids. IEEE Journal of Emerging and Selected Topics in Power Electronics. 2021;**9**(2): 1205-1221. DOI: 10.1109/JESTPE.2020.2978064

[48] Hossain MZ, Rahim NA, Selvaraj I. Recent progress and development on power DC-DC converter topology, control, design and applications: A review. Renewable and Sustainable Energy Reviews. 2018;**81**:205-230. DOI: 10.1016/j.rser.2017.07.017

[49] Sivakumar S, Sathik MJ, Manoj PS, Sundararajan G. An assessment on performance of DC–DC converters for renewable energy applications. Renewable and Sustainable Energy Reviews. 2016;**58**:1475-1485. DOI: 10.1016/j.rser.2015.12.057

[50] Amir A, Amir A, Che HS, Elkhateb A, Rahim NA. Comparative analysis of high voltage gain DC-DC converter topologies for photovoltaic systems. Renewable Energy. 2019;**136**: 1147-1163. DOI: 10.1016/j.renene.2018.09.089

[51] Raghavendra KVG, Zeb K, Muthusamy A, Krishna TNV, Kumar SVSVP, Kim D-H, et al. A

comprehensive review of DC–DC converter topologies and modulation strategies with recent advances in solar photovoltaic systems. Electronics. IEEE. 2020;**9**(1):31. DOI: 10.3390/electronics 9010031

[52] Nikhar AR, Apte SM, Somalwar R. Review of various control techniques for DC-DC interleaved boost converters. In: 2016 International Conference on Global Trends in Signal Processing, Information Computing and Communication (ICGTSPICC). Jalgaon, India: IEEE; 2016. pp. 432-437. DOI: 10.1109/ICGTSPICC.2016.7955340

[53] Reshma Gopi R, Sreejith S. Converter topologies in photovoltaic applications – A review. Renewable and Sustainable Energy Reviews. 2018;**94**:1-14. DOI: 10.1016/j.rser.2018.05.047

[54] Poorali B, Torkan A, Adib E. High step-up Z-source DC–DC converter with coupled inductors and switched capacitor cell. IET Power Electronics. 2015;**8**(8):1394-1402. DOI: 10.1049/iet-pel.2014.0200

[55] Dharmasena S, Olowu TO, Sarwat AI. Bidirectional AC/DC converter topologies: A review. 2019 SoutheastCon. 2019:1-5. DOI: 10.1109/SoutheastCon42311.2019.9020287

[56] Jotham Jeremy L, Ooi CA, Teh J. Non-isolated conventional DC-DC converter comparison for a photovoltaic system: A review. Journal of Renewable and Sustainable Energy, Energy. 2020; **12**(1):13502. DOI: 10.1063/1.5095811

[57] Siddharthan N, Balasubramanian B. Performance evaluation of SEPIC, Luo and ZETA converter. International Journal of Power Electronics and Drive Systems. 2019;**10**(1):374. DOI: 10.11591/ijpeds.v10.i1.pp374-380

[58] Forouzesh M, Shen Y, Yari K, Siwakoti YP, Blaabjerg F. High-efficiency high step-Up DC–DC converter with dual coupled inductors for grid-connected photovoltaic systems. IEEE Transactions on Power Electronics. 2018;**33**(7):5967-5982. DOI: 10.1109/TPEL.2017.2746750

[59] Hu X, Gong C. A high gain input-parallel output-series DC/DC converter with dual coupled inductors. IEEE Transactions on Power Electronics. 2015; **30**(3):1306-1317. DOI: 10.1109/TPEL.2014.2315613

[60] Sagar Bhaskar M, Meraj M, Iqbal A, Padmanaban S, Kiran Maroti P, Alammari R. High Gain Transformer-Less Double-Duty-Triple-Mode DC/DC Converter for DC Microgrid. IEEE Access. 2019;7:36353-36370. DOI: 10.1109/ACCESS.2019.2902440

[61] Qais MH, Hasanien HM, Alghuwainem S. Enhanced whale optimization algorithm for maximum power point tracking of variable-speed wind generators. Applied Soft Computing. 2020;**86**:105937. DOI: 10.1016/j.asoc.2019.105937

[62] Qais M, Hasanien HM, Alghuwainem S. Salp swarm algorithm-based TS-FLCs for MPPT and fault ride-through capability enhancement of wind generators. ISA Transactions. 2020;**101**: 211-224. DOI: 10.1016/j.isatra.2020.01.018

[63] Mahmoud HY, Hasanien HM, Besheer AH, Abdelaziz AY. Hybrid cuckoo search algorithm and grey wolf optimiser-based optimal control strategy for performance enhancement of HVDC-based offshore wind farms. IET Generation, Transmission & Distribution. 2020;**14**(10):1902-1911. DOI: 10.1049/iet-gtd.2019.0801

[64] Adnan M, Oninda M, Nishat M, Islam N. Design and simulation of a DC-DC boost converter with PID controller for enhanced performance. International Journal of Engineering Research & Technology. 2017;**6**:27-32. DOI: 10.17577/IJERTV6IS090029

[65] Hekimoğlu B, Ekinci S, Kaya S. Optimal PID controller design of DC-DC buck converter using whale optimization algorithm. In: 2018 International Conference on Artificial Intelligence and Data Processing (IDAP). Malatya, Turkey: IEEE; 2018. pp. 1-6. DOI: 10.1109/IDAP.2018.8620833

[66] Utkin V, Poznyak A, Orlov Y, Polyakov A. Conventional and high order sliding mode control. Journal of the Franklin Institute. 2020;**357**(15): 10244-10261. DOI: 10.1016/j.jfranklin.2020.06.018

[67] Mohd Zaihidee F, Mekhilef S, Mubin M. Robust speed control of PMSM using sliding mode control (SMC)—A review. Energies. 2019;**12**(9). DOI: 10.3390/en12091669

[68] Utkin V, Lee H. Chattering problem in sliding mode control systems. In: International Workshop on Variable Structure Systems. VSS'06, 2006. Alghero, Sardinia: IEEE; 2006. pp. 346-350. DOI: 10.1109/VSS.2006.1644542

[69] Wei Q, Wu B, Xu D, Zargari NR. Model predictive control of capacitor voltage balancing for cascaded modular DC–DC converters. IEEE Transactions on Power Electronics. 2017;**32**(1): 752-761. DOI: 10.1109/TPEL.2016.2530869

[70] An F, Song W, Yang K, Hou N, Ma J. Improved dynamic performance of dual active bridge dc–dc converters using MPC scheme. IET Power Electronics. 2018;**11**(11):1756-1765. DOI: 10.1049/iet-pel.2017.0707

[71] Irmak E, Güler N. A model predictive control-based hybrid MPPT method for boost converters. International Journal of Electronics. 2020;**107**(1):1-16. DOI: 10.1080/00207217.2019.1582715

[72] Tahir S, Wang J, Baloch MH, Kaloi GS. Digital control techniques based on voltage source inverters in renewable energy applications: A review. Electronics. 2018;**7**(2). DOI: 10.3390/electronics7020018

[73] Azer P, Emadi A. Generalized state space average model for multi-phase interleaved buck, boost and buck-boost DC-DC converters: Transient, steady-state and switching dynamics. IEEE Access. 2020;**8**:77735-77745. DOI: 10.1109/ACCESS.2020.2987277

[74] Tarbosh QA et al. Review and investigation of simplified rules fuzzy logic speed controller of high performance induction motor drives. IEEE Access. 2020;**8**:49377-49394. DOI: 10.1109/ACCESS.2020.2977115

[75] Soliman MA, Hasanien HM, Azazi HZ, El-Kholy EE, Mahmoud SA. An adaptive fuzzy logic control strategy for performance enhancement of a grid-connected PMSG-based wind turbine. IEEE Transactions on Industrial Informatics. 2019;**15**(6):3163-3173. DOI: 10.1109/TII.2018.2875922

[76] Silva JF, Pinto SF. "35 - Linear and nonlinear control of switching power converters," In: Rashid MH editors. Power Electronics Handbook 4 ed. Butterworth-Heinemann. 2018. pp. 1141-1220

[77] Alhejji A, Mosaad MI. Performance enhancement of grid-connected PV

systems using adaptive reference PI controller. Ain Shams Engineering Journal. 2021;**12**(1):541-554. DOI: 10.1016/j.asej.2020.08.006

[78] Mostafa MR, Saad NH, El-sattar AA. Tracking the maximum power point of PV array by sliding mode control method. Ain Shams Engineering Journal. 2020;**11**(1):119-131. DOI: 10.1016/j.asej.2019.09.003

[79] Yfoulis C, Papadopoulou S, Voutetakis S. Enhanced control of a buck-boost DC-DC converter via a closed-form MPC reference governor scheme. IECON 2019 - 45th Annual Conference of the IEEE Industrial Electronics Society. 2019;**1**: 365-370. DOI: 10.1109/IECON.2019.8927486

[80] Ibrahim O, Yahaya NZ, Saad N. State-space modelling and digital controller design for DC-DC converter. TELKOMNIKA (Telecommunication, Computing, Electronics and Control). 2016;**14**:497-506. DOI: 10.12928/TELKOMNIKA.v14i1.3042

[81] Freytes J, Bergna G, Are Suul J, D'Arco S, Saad H, Guillaud X. "State-space modelling with steady-state time invariant representation of energy based controllers for modular multilevel converters," in 2017 IEEE Manchester PowerTech. 2017. pp. 1–7. DOI: 10.1109/PTC.2017.7981011.

[82] Hassan SA, Iqbal S. Automatic car braking system using fuzzy logic controller with environmental factors. In: 2019 22nd International Multitopic Conference (INMIC). Islamabad, Pakistan: IEEE; 2019. pp. 1-8. DOI: 10.1109/INMIC48123.2019.9022773

[83] Ishaque MR et al. Fuzzy logic-based duty cycle controller for the energy management system of hybrid electric vehicles with hybrid energy storage system. Applied Sciences. 2021;**11**(7). DOI: 10.3390/app11073192

[84] Xiang X, Yu C, Lapierre L, Zhang J, Zhang Q. Survey on fuzzy-logic-based guidance and control of marine surface vehicles and underwater vehicles. International Journal of Fuzzy Systems. 2018;**20**(2):572-586. DOI: 10.1007/s40815-017-0401-3

[85] Loukil K, Abbes H, Abid H, Abid M, Toumi A. Design and implementation of reconfigurable MPPT fuzzy controller for photovoltaic systems. Ain Shams Engineering Journal. 2020;**11**(2):319-328. DOI: 10.1016/j.asej.2019.10.002

[86] Velayudhan AKD. Design of a supervisory fuzzy logic controller for monitoring the inflow and purging of gas through lift bags for a safe and viable salvaging operation. Ocean Engineering. 2019;**171**:193-201. DOI: 10.1016/j.oceaneng.2018.10.049

[87] Hema Rani P, Navasree S, George S, Ashok S. Fuzzy logic supervisory controller for multi-input non-isolated DC to DC converter connected to DC grid. International Journal of Electrical Power & Energy Systems. 2019;**112**: 49-60. DOI: 10.1016/j.ijepes.2019.04.018

[88] Oudda M, Hazzab A. Photovoltaic system with SEPIC converter controlled by the fuzzy logic. International Journal of Power Electronics and Drive Systems. 2016;7:1283. DOI: 10.11591/ijpeds.v7.i4.pp1283-1293

[89] Eltawil MA, Zhao Z. MPPT techniques for photovoltaic applications. Renewable and Sustainable Energy Reviews. 2013;**25**(C):793-813. Available from: https://econpapers.repec.org/RePEc:eee:rensus:v:25:y:2013:i:c:p

[90] Niasse OA et al. Optimization of electrics poarameters CdS/CdTe thin film solar cell using dielectric model. World Journal of Condensed Matter Physics. 2016;**06**(02):75-86. DOI: 10.4236/wjcmp.2016.62011

[91] Quoc DP et al. The new combined maximum power point tracking algorithm using fractional estimation in photovoltaic systems. In: 2011 IEEE Ninth International Conference on Power Electronics and Drive Systems. Singapore: IEEE; 2011. pp. 919-923. DOI: 10.1109/PEDS.2011.6147363

[92] Noguchi T, Togashi S, Nakamoto R. Short-current pulse-based maximum-power-point tracking method for multiple photovoltaic-and-converter module system. IEEE Transactions on Industrial Electronics. 2002;**49**(1): 217-223. DOI: 10.1109/41.982265

[93] Chekired F, Larbes C, Rekioua D, Haddad F. Implementation of a MPPT fuzzy controller for photovoltaic systems on FPGA circuit. Energy Procedia. 2011; **6**:541-549. DOI: 10.1016/j.egypro.2011.05.062

[94] Radjai T, Rahmani L, Mekhilef S, Gaubert JP. Implementation of a modified incremental conductance MPPT algorithm with direct control based on a fuzzy duty cycle change estimator using dSPACE. Solar Energy. 2014;**110**:325-337. DOI: 10.1016/j.solener.2014.09.014

[95] Femia N, Petrone G, Spagnuolo G, Vitelli M. A technique for improving P&O MPPT performances of double-stage grid-connected photovoltaic systems. IEEE Transactions on Industrial Electronics. 2009;**56**(11):4473-4482. DOI: 10.1109/TIE.2009.2029589

[96] Huang L, Zhang J, Cui Q, Wang H, Shu J. Maximum power point tracking scheme with partial shading detection for two-stage grid-connected photovoltaic inverters. Journal of Engineering. 2019;**2019**(16):2556-2562. DOI: 10.1049/joe.2018.8547

[97] Elgendy MA, Zahawi B, Atkinson DJ. Assessment of perturb and observe MPPT algorithm implementation techniques for PV pumping applications. IEEE Transactions on Sustainable Energy. 2012;**3**(1):21-33. DOI: 10.1109/TSTE.2011.2168245

[98] Suwannatrai P, Liutanakul P, Wipasuramonton P. Maximum power point tracking by incremental conductance method for photovoltaic systems with phase shifted full-bridge dc-dc converter. In: The 8th Electrical Engineering/Electronics, Computer, Telecommunications and Information Technology (ECTI) Association of Thailand - Conference, 2011. Khon Kaen, Thailand: IEEE; 2011. pp. 637-640. DOI: 10.1109/ECTICON.2011.5947920

[99] Selvan D. Modeling and simulation of incremental conductance MPPT algorithm for photovoltaic applications. International Journal of Engineering, Science and Technology. 2013;**2**:681-685

[100] Mei Q, Shan M, Liu L, Guerrero JM. A novel improved variable step-size incremental-resistance MPPT method for PV systems. In: IEEE Transactions on Industrial Electronics. Vol. 58, No. 6. June 2011. pp. 2427-2434. DOI: 10.1109/TIE.2010.2064275

Chapter 2

A Study on Fiber Optic Temperature Sensor Using Al_2O_3 as High Index Overlay for Solar Cell Applications

Subramaniyam Narasimman, Lakshmi Narayanan Balakrishnan, Arunkumar Chandrasekhar and Zachariah C. Alex

Abstract

Recently, the performance of solar cell is impacted by rising panel temperatures. For solar cells to work at their best and have the longest possible useful life, the temperature of the panels must be kept at an ideal level. Current temperature sensors have a slow response time, poor accuracy, and low resolution. Meanwhile, Al_2O_3 and its derivatives have demonstrated a noteworthy role in temperature sensing applications due to its greater surface area, ease of synthesis, tailored optical characteristics, high melting point, and high thermal expansion coefficient. Al_2O_3-based nanoparticles have been employed in fiber optic-based temperature sensors as a sensing layer, a sensitivity improvement material, and a sensing matrix material. In this chapter, we discuss the function of Al_2O_3-based nanomaterials in evanescent wave-based temperature sensors, sensing characteristics such as sensitivity, linearity, and repeatability. The ZAZ-based sensor (Section 3.1) shows an operating temperature range between 100.9°C and 1111.0°C, the temperature sensitivity becomes 1.8×10^{-5}/°C. The fabricated sensor had a linearity of 99.79%. The synthesized Al_2O_3 nanoparticles (Section 3.2) were given better linearity and high sensitivity (~27) at 697 nm compared with other sensing materials such as ZnO, SnO_2, TiO_2. The Al_2O_3-MgO (50–50%) (Section 3.3) demonstrated an ultrahigh sensitivity of 0.62%/°C with a better linear regression coefficient of 95%. The present advances and problems are also discussed in detail.

Keywords: Al_2O_3, fiber optic sensor, clad modification technology, temperature sensor, solar cell

1. Introduction

Solar energy is one of the more well-known renewable energy sources, and businesses and industries use its energy harvesting techniques extensively. However, one significant disadvantage of commercial solar cells is their low efficiency at higher panel temperatures. The panel temperature, solar radiation, shading, panel inclination, alignment, dust, and maintenance have a significant impact on the energy efficiency of solar cells.

IntechOpen

A one-degree temperature rise can reduce the efficiency by ~0.045% over a temperature range of 15–60°C in a monocrystalline silicon solar cell [1]. Till now more studies on the use of Al_2O_3 in solar cells existed. For instance, El-Shafai et al. have prepared a novel hybrid nanomaterial (HNM) (GO@CuO.γ-Al_2O_3) and studied their thermal and electrical performances. Different nanofluids were prepared from mono NMs (GO, CuO, and γ-Al_2O_3), and hybrid NMs (GO@CuO, GO@γ-Al_2O_3, and GO@CuO.γ-Al_2O_3) with water as a base fluid, to study the thermal conductivity. Different concentrations of the nanofluids (0.0625, 0.125, and 0.2%) were investigated within a temperature range of 20–50°C. Compared to water, GO@CuO.γ-Al_2O_3 shown maximum enhancement in thermal conductivity (22.56%) with 0.2% concentration and 50°C which is favorable for solar collector heaters. Gunjo et al. have reported that adding 5% of Al_2O_3 to paraffin-based nanofluids improved the melting rate by 3.46 times and discharge rate by 3 times compared with pure paraffin-based nanofluids. This increases thermal conductivity, dynamic viscosity, and density and also lowers the heat storage compared with pure paraffin-based nanofluids which favors solar energy applications. Khalifa et al. have prepared colloidal Al_2O_3 nanoparticles by electrolysis method and deposited over p-type silicon wafer by drop casting method and investigated solar cell performances. Mahmoud et al. have studied the performances aluminum oxide (α- Al_2O_3) nanoparticle and metal aluminate spinel nanoparticle (M- Al_2O_4, where M is Co, Cu, Ni, Zn) as photo-anodes in quantum dot photovoltaic cells. Electrochemical impedance spectroscopy shows that Zn Al_2O_4 and Ni Al_2O_4 nanocomposites have the highest lifetimes of the photogenerated electrons (τ_n) of $11*10^{-2}$ and $96*10^{-3}$ ms, respectively, and the lowest diffusion rates (Keff) of 9.09 and 10.42 ms^{-1}, respectively. Amalraj et al. have reported Al_2O_3 nanoparticles as coolant materials show good efficiency in solar cooling panels. The solar cooling efficiency is 12 and the fill factor is 0.55. However, due to electromagnetic, chemical, and mechanical disturbances, the traditional NTC and PTC-type thermistor, thermocouples, and resistance temperature detectors (RTDs) [2] are unable to give adequate performance for certain real-time applications. Thus, fiber optic sensors would be a superior choice for these applications as the optical signal is immune to electromagnetic field interference, can be used for long-distance communication with low loss, and is compact and simple to employ in real-time applications [3].

Currently, a variety of fiber optic temperature sensors have been reported, including Fabry-Perot, Mach-Zehnder, Fiber Bragg grating, thin film on fiber core, microfiber and coating nanoscale level sensing layer at cladding modified fiber (CMF) and a fiber's tip [4–15]. Of all, the CMF-based sensors outperform existing ones in all ways, including less weight, electromagnetic interference resistance, ease of manufacture, and increased accuracy in challenging conditions [16]. In CMF, a slight change in the modified nanomaterial due to temperature resulted in light intensity variation. For instance, Huang et al. [15] have reported $ZrO_2/Al_2O_3/ZrO_2$ (ZAZ) coated fiber optic temperature sensor and achieved better sensitivity of 1.8×10^{-5}/°C. According to Sun et al. [17], an optical fiber temperature sensor based on temperature cross sensitivity with a RI sensitive device had an improved sensitivity of 350 pm/°C. Rahman et al. [18] have demonstrated bimetallic layer fiber optic temperature sensor and showed moderate temperature sensitivity. Nevertheless, these sensors have their own limitations, such as design complexity, low resolution, poor sensitivity, non-linearity, and dynamic range. By SMO coated CMF, these issues can be resolved.

Till now, a multitude of SMOs has been used in the construction of fiber optic sensors, including ZnO [19], TiO_2 [20], SnO_2 [21], Al_2O_3 [22], MgO [23], and SiO_2 [24]. A substantial increase in temperature sensor research for improved linearity and sensitivity development has been observed in the last 10 years. Because of their numerous

Material	Density, ρ (Kg/cm^3)	Specific heat, C_p (J/Kg K)	Thermal conductivity, k (W/m K)	Viscosity, μ (Kg/ms)
Al_2O_3	981	4189	0.643	0.0006

Table 1.
Thermophysical properties of Al_2O_3.

features, nanomaterials have been used in the construction of temperature sensors, greatly advancing the field of temperature sensor research. Due to their exceptional qualities, including a high melting point, a high thermal expansion coefficient, and good physical, chemical, and optical properties, Al_2O_3 and its nanocomposites have revolutionized the world of sensing. The thermophysical properties (Thermal Conductivity, Viscosity, Density, Specific heat) of Al_2O_3 nanoparticle were shown in **Table 1**. In temperature sensors Al_2O_3-based nanomaterials have been used as a sensing layer, to provide a large surface area and compatibility for temperature detection. In this review, we aim to thoroughly outline the role of Al_2O_3-based nanomaterials in temperature-based sensors, their present advancements, and challenges.

2. Polymer-based fiber optic sensors (PFOS)

Recently, modern technological fields like structural, aeronautical, and aerospace engineering, as automotive, industrial, and medical engineering all heavily rely on sensors. However, till now most of the sensors are based on electrical transduction mechanism. Besides that, sensors based on optics facilitate multitude of benefits such as flexibility, anti-electromagnetic interference, low cost, and easy fabrication compared with conventional electrical sensors. The use of polymer-based fibers in the manufacture of sensors allows the researcher to create any type of fiber geometries, which is a problem with glass fiber that has not yet been resolved. Another advantage of polymer-based fiber optic sensors is the ease with which they can be modified by imprinted polymers. Materials are also another important key parameter for polymer-based optical fiber fabrication and it can be divided into two categories: plastic and natural materials. Poly(methyl methacrylate) (PMMA), cyclo-olefin polymer (COP, ZeonexTM), polycarbonate (PC), amorphous fluoropolymer (CYTOPTM), and PDMS (polydimethylsiloxane) are the most common materials for polymer-based optical fiber fabrications (Chemosensor). The fiber can be made from a single material or a combination of polymer materials when cladding and core are made from different materials. Generally, these polymers-based optical fibers provide good optical and mechanical properties, ease of access, and use [25]. POF can be produced as a single-mode [26, 27] and multimode fiber [28, 29], with a step [30] or gradient index refractive index profile [31]. The fiber has opened up a broad number of uses depending on the type of fiber, diameter of the core/cladding, refractive index of the core/cladding, numerical aperture, and dopants utilized in the fabrication process. **Figure 1** shows the record of research articles in the database Scopus.

2.1 POF temperature sensing

Basically, the fiber optic sensor consists of the light source, sensor element, and detector which can be further interfaced with the data acquisition device (**Figure 2**). An optical fiber (single-mode or multimode) is used to direct a light source, such as

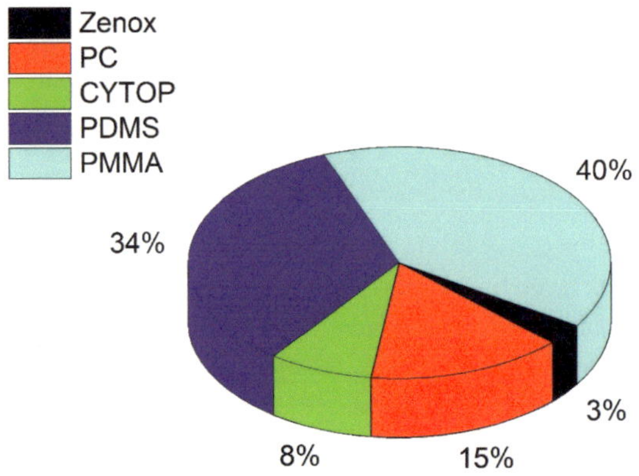

Figure 1.
The percentage of a given polymer in the total number of articles on POF published in 2015–2022 [32].

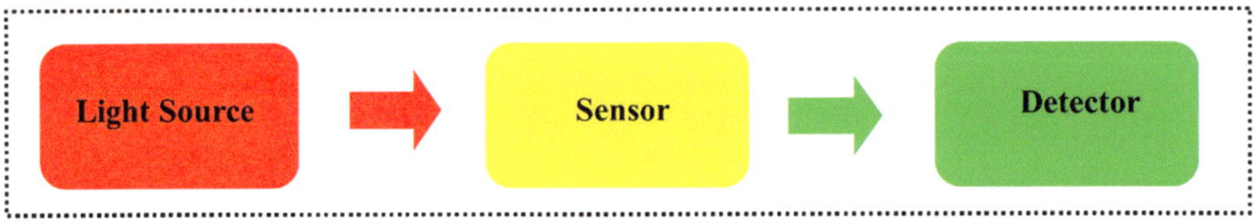

Figure 2.
The basic setup of the fiber optic sensor.

a laser (narrowband source), LED (broadband source) to the spectrometer, where optical signal variations will be the measurable quantity of interest (Temperature). The temporal or spectral domains can be used to analyze the measurement signal. A photodiode and an optical spectrum analyzer or a spectrometer can be used as the detection system [33]. Generally, POF sensors offer distributed [34] and pointwise sensing measurements. However, owing to high attenuation, the POF sensors are dedicated mainly to pointwise measurement. Temperature sensing by the POF sensor can be an interaction of temperature parameter that causes changes in the intensity (amplitude), frequency, phase, and polarization of the transmitted light [35–37].

2.2 Evanescent wave absorption in optical fiber

Typically, synthesized nanoparticles (η_{Al2O3} = 1.763) were used in the CMF sensor to replace the natural cladding (η_{clad} = 1.402). As a result, the sensor enters a leaky mode known as η_{mclad}> > η_{core}, which results in a reduction in propagated light intensity. The evanescent absorption in the modified cladding's changing refractive index could be the cause of the output light's temperature-dependent intensity variation. When the light was guided through the fiber under total internal reflection at the core/modified cladding interface, a portion of the light was transmitted into the modified cladding region, where a portion will be reentered back for propagation based on the change in the modified cladding's refractive index and the rest will be lost. The intensity of this phenomenon, known as an evanescent field, decreases exponentially the farther it is from the surface [38]. When atmosphere temperature varies, the light intensity that travels through the fiber changes as a result of changes in modified cladding refractive index which affects sensor output. The crux of temperature sensing is mainly due to changes in refractive index because of the thermal expansion and thermo-optic effect [39, 40].

2.3 Al_2O_3 nanomaterials for temperature sensing

The development of nanotechnology over the past 10 years has greatly sparked interest in every area of science and technology. In order to create novel materials, nanotechnology has been primarily used to reorganize bulk materials at the nanoscale. Al_2O_3-based nanoparticles, among other forms of nanomaterials, have drawn a lot of interest from the scientific community because of their exceptional qualities, including a high melting point, a high thermal expansion coefficient, and good physical, chemical, and optical properties. Both amorphous and crystalline phases can be found in aluminum oxide, often known as alumina (Al_2O_3). The crystalline form of Al_2O_3 has a number of metastable structural phases, including corundum, which is a stable phase with a rhombohedral structure, as well as the following: Al_2O_3 (monoclinic structure), Al_2O_3 (orthorhombic structure), Al_2O_3 (cubic or hexagonal structure), Al_2O_3 (tetragonal structure), and Al_2O_3 (orthoclinic structure). Aluminum (Al^{3+}) ions and vacancies are randomly distributed throughout tetrahedral (AlO_4), polyhedral (AlO_5), and octahedral (AlO_6) sites in amorphous Al_2O_3 [41, 42]. Numerous papers describe the use of Al_2O_3 and its composites in applications for gas sensing, environmental analysis, and temperature sensing.

2.4 Synthesis of Al_2O_3 nanomaterials

The synthesis of Al_2O_3 is done by various approaches such as sol-gel, PVD, CVD, hydrothermal, co-precipitation, solvothermal or sonochemical methods [43, 44]. Co-precipitation is one of these methods, and because of its capacity to produce large quantities of Al_2O_3 nanostructure at a low cost and with no environmental impact, it has the potential for application in this process. Co-precipitation was employed to synthesize Al_2O_3 nanoparticles. In a nutshell, 50 ml of distilled water was mixed with 2.12 g (0.2 M) of Al $(NO_3)_2$. $9H_2O$ precursor, and the mixture was vigorously agitated for 2 hours. Drop by drop, 10 ml of 2 M NaOH stock solution was added to the previously combined solution while being constantly stirred until the pH level reached 8.0. The reacted solution was additionally left at room temperature for 24 hours. The white precipitate that resulted was then washed three times. The resultant nanoparticles were filtered and calcinated at 600°C for 4 h.

2.5 Sensor region preparation

Figure 3 depicts the schematic diagram of the metal oxide-coated temperature sensing system. The transmitted light spectra of our proposed sensor were studied using a broadband light source (SLS201/M) and an optical spectrometer. The cladding modification method was used to achieve a fiber optic temperature sensor probe (CMM). A central portion of a PMMA optical fiber was denuded and etched with an acetone solution. The synthesized metal oxide nanoparticles were mixed with double distilled water to form a paste, which was then deposited in a dip coating technique to the etched surface to a thickness of 20 μm (**Figure 4**). The coated optical fiber was then allowed to dry at ambient temperature and employed as a sensing region. The sensor was kept in a temperature-sensing chamber, and a heater coupled with microcontroller was used to regulate the temperature. The holder was used to clip the two ends of the fiber, preventing interference from outside disturbances. Glue was used to firmly seal the sensor chamber [46].

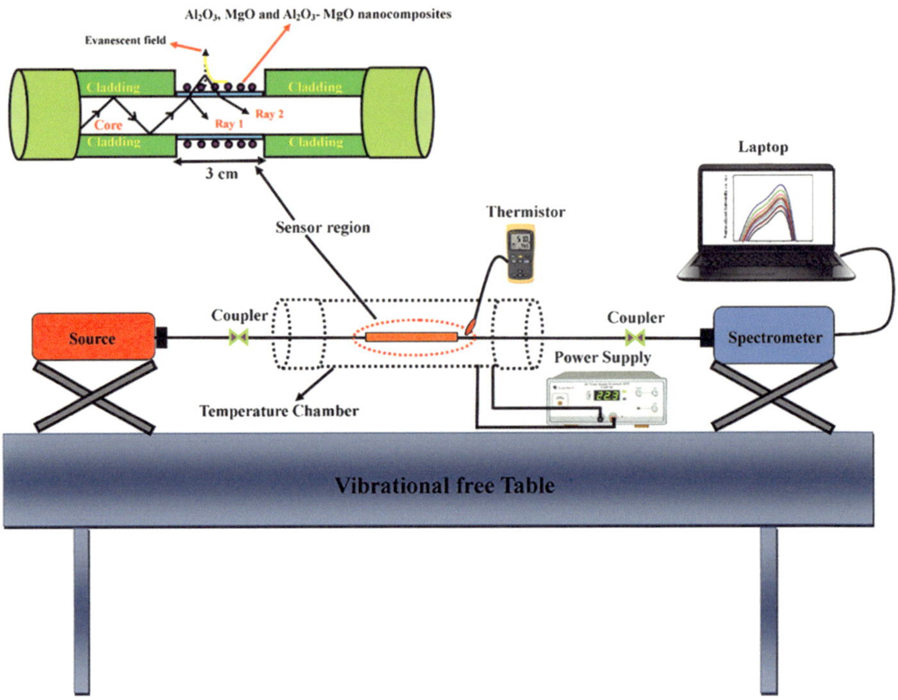

Figure 3.
Schematic diagram of the fiber optic temperature sensor setup [45].

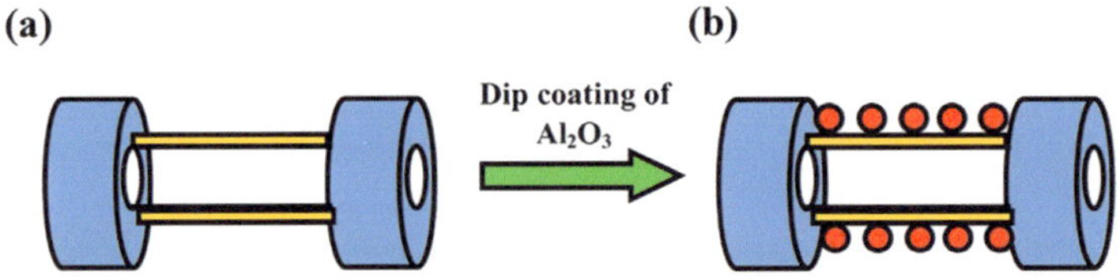

Figure 4.
Scheme of probe fabrication (a) after etching (b) after coating with metal oxides as sensing layer.

3. Assessment of temperature sensing performances

3.1 Al_2O_3 nanomaterials-based temperature sensor

In this chapter, the sensor head was built with three layers of $ZrO_2/Al_2O_3/ZrO_2$ (ZAZ) dielectric materials via physical vapor deposition (PVD) onto the tip of a sapphire fiber [15]. The sensor was held within a high-temperature furnace for temperature sensing following ZAZ deposition and thermal annealing. As seen in **Figure 5**, a 3 dB multimode optical fiber coupler, a k-type thermocouple, a broad-band light source, and a fiber optic spectrometer were all used in the measurement. Due to the thermo-optic and elastic-optic effects, the refractive index and thickness of the ZAZ films will rise as the ambient temperature rises.

This will cause the interference spectra to shift and the optical path difference (OPD) of the thin film interferometer to vary. Additionally, a rise in the ambient temperature will cause the interference spectra to shift. As a result, it is possible to

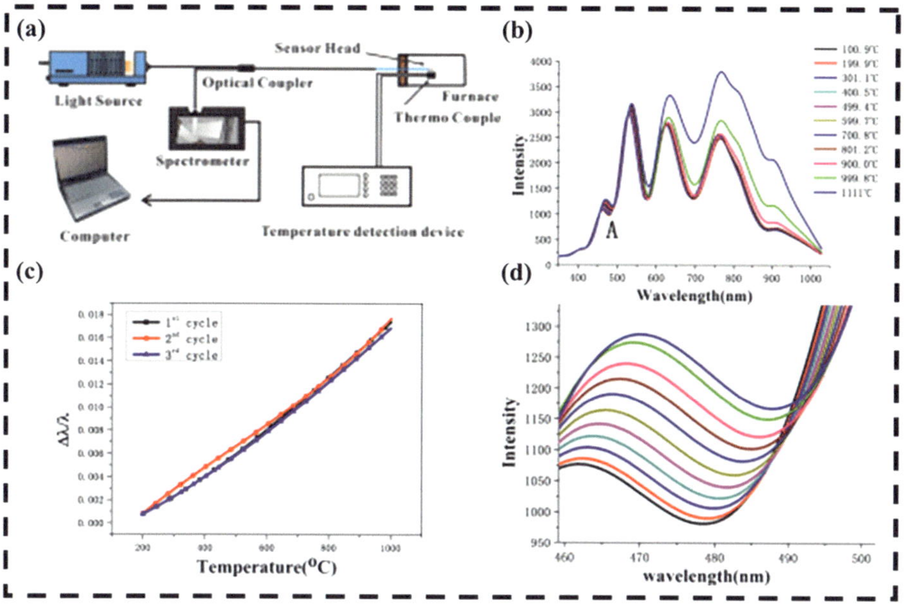

Figure 5.
(a) Optical interrogation system for measurement of temperature. (b) Reflection spectra of the thin-film sensor at different temperatures. (c) the variation in OPD under different temperatures of two thermal cycles. (d) Enlargement of reflection spectra around 480 nm [15].

measure the ambient temperature. When the ambient temperature changes from 100.9 to 1111.0°C, the temperature sensitivity becomes 1.8×10^{-5}/°C. The fabricated sensor had a linearity of 99.79%.

3.2 Fabrication of fiber optic-based temperature sensor with various metal oxides (ZnO, SnO_2, Al_2O_3, and TiO_2) as sensing layer

The co-precipitation approach was used in this study to synthesize metal oxide semiconductors (ZnO, SnO_2, Al_2O_3, and TiO_2), which were then subjected to several material characterization techniques [47]. According to the XRD data, the ZnO nanoparticle crystallized in a hexagonal wurtzite structure, while SnO_2 nanoparticles are in a rutile tetragonal structure, Al_2O_3 nanoparticles are in a rhombohedral structure, and TiO_2 nanoparticles are in a rutile anatase structure. The SEM analysis confirms that all of the synthesized nanopowders are in grains that are dispersed equally (**Figure 6**). Additionally, dip coating was used to deposit the synthesized metal oxide semiconductors over the optical fiber's cladding-modified region, and investigated temperature sensing for broad wavelength range and specific wavelength ranges (Blue, Green, Orange, Red, and Yellow). **Figure 7** depicts the change in light intensity of metal oxide nanoparticles (ZnO, SnO_2, Al_2O_3, and TiO_2) at various temperatures between 35 and 75°C with a 5°C step interval.

The three characteristic peaks appear in the spectrum at wavelengths of 697, 774, and 952 nm respectively which are the characteristic spectrum of the optical fiber used. The characteristic spectrum shows intensity variations at various temperatures. In comparison to the other two characteristic peaks, the temperature variation led to the greatest peak intensity variation at about 697 nm. **Figure 8** depicts the variation in light intensity of Al_2O_3 nanoparticles at various temperatures, from 35 to 75°C with a step

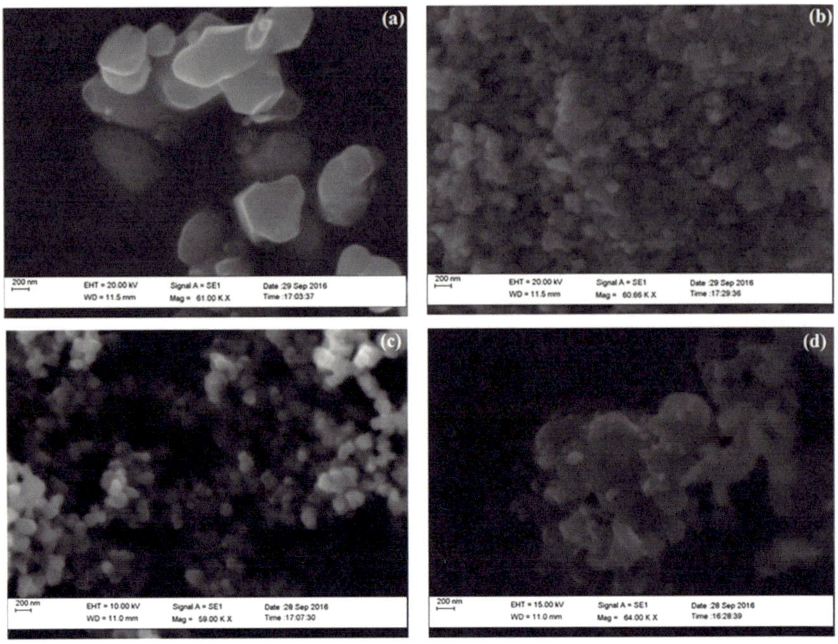

Figure 6.
SEM micrographs of (a) ZnO, (b) SnO_2, (c) TiO_2, and (d) Al_2O_3 nanopowders [47].

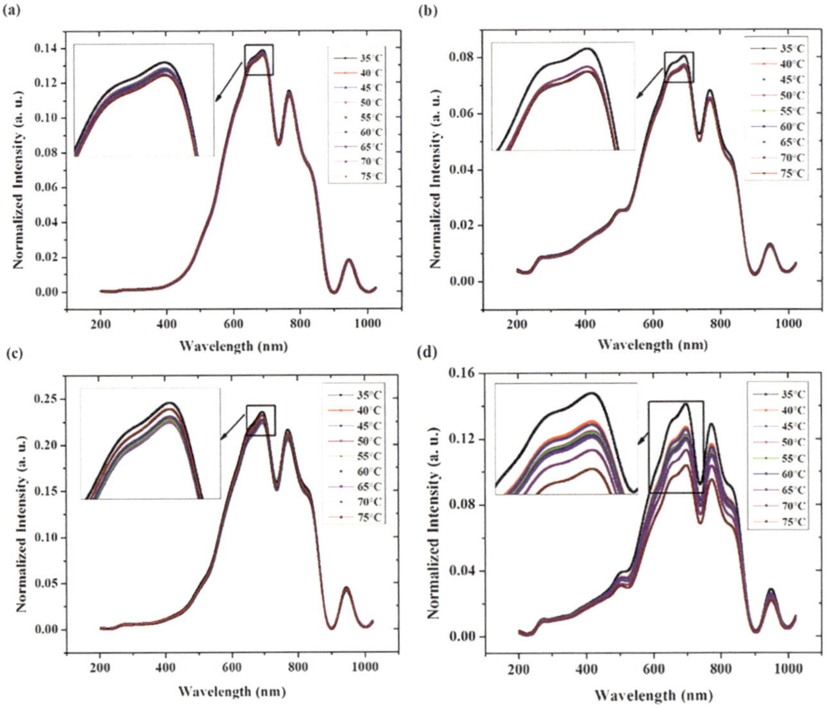

Figure 7.
Spectral response of synthesized nanopowders (a) ZnO (b) SnO_2 (c) TiO_2 and (d) Al_2O_3 for various temperature (35 to 75°C) at 697 nm [47].

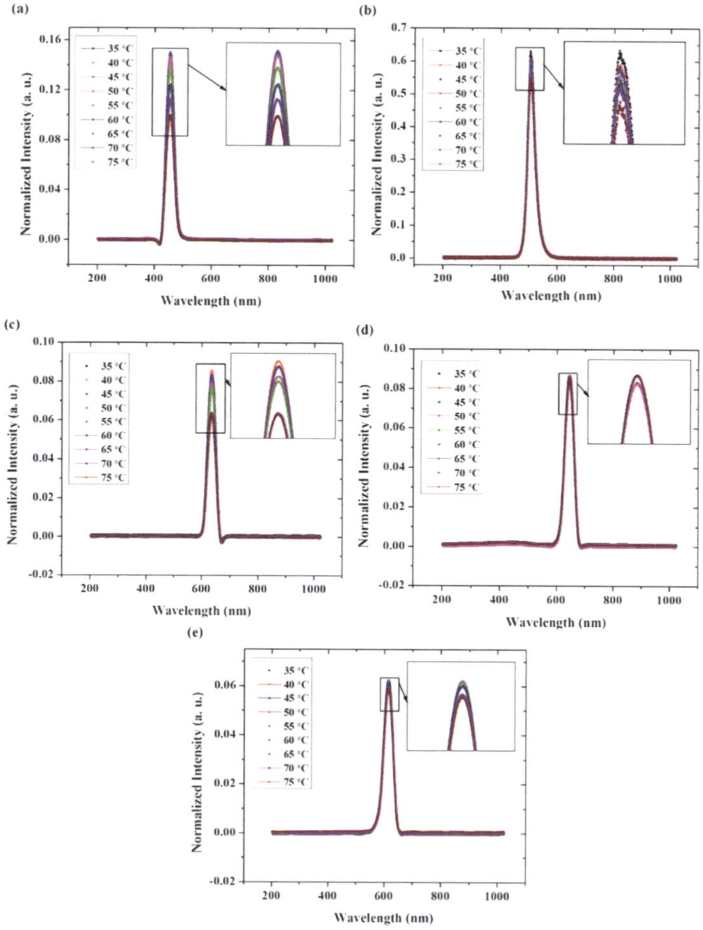

Figure 8.
Spectral response of Al_2O_3 nanopowders for various temperature (35 to 75°C) at different wavelength ranges (a) blue, (b) green, (c) orange, (d) red, and (e) yellow [47].

interval of 5°C (Blue, Green, Orange, Red, and Yellow). When compared to other wavelengths, it has been shown that blue and orange wavelengths displayed a significant variation in intensity. It shows that both blue and orange wavelength ranges are possible for the manufactured fiber optic temperature sensor to operate in. It reveals that the synthesized Al_2O_3 nanoparticles were given better linearity and high sensitivity (~27) at 697 nm compared with other sensing materials. Further, wavelength dependent temperature sensing characteristics of Al_2O_3 nanopowders were studied and it shows better sensitivity (~34) in the blue wavelength region (450 nm–495 nm) (**Figure 9**).

3.3 Fabrication of fiber optic-based temperature sensor with Al_2O_3, MgO, and composites as sensing layer

Fabrication and characterization of fiber optic temperature sensors using Al_2O_3-MgO nanocomposite as cladding material have been reported [45]. To synthesize Al_2O_3, MgO, and their various compositions, the co-precipitation technique was chosen and subjected to various material characterizations. From, XRD, Al_2O_3 and

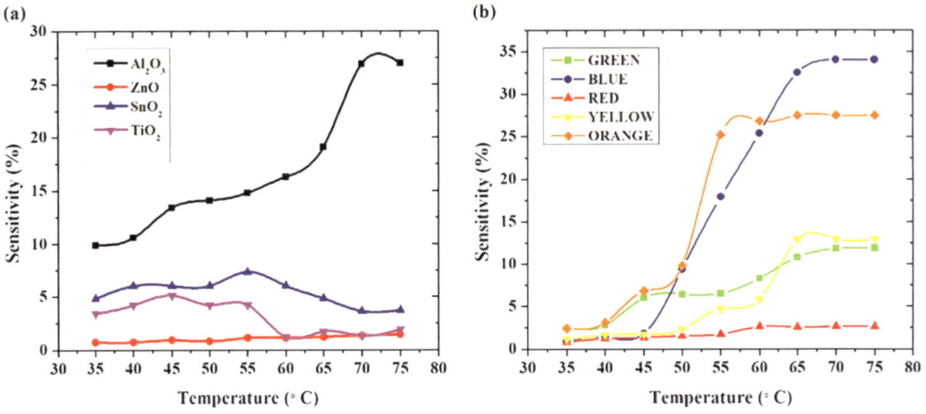

Figure 9.
Temperature sensitivity of (a) synthesized nanopowders at 697 nm and (b) Al_2O_3 nanopowder at different wavelengths [47].

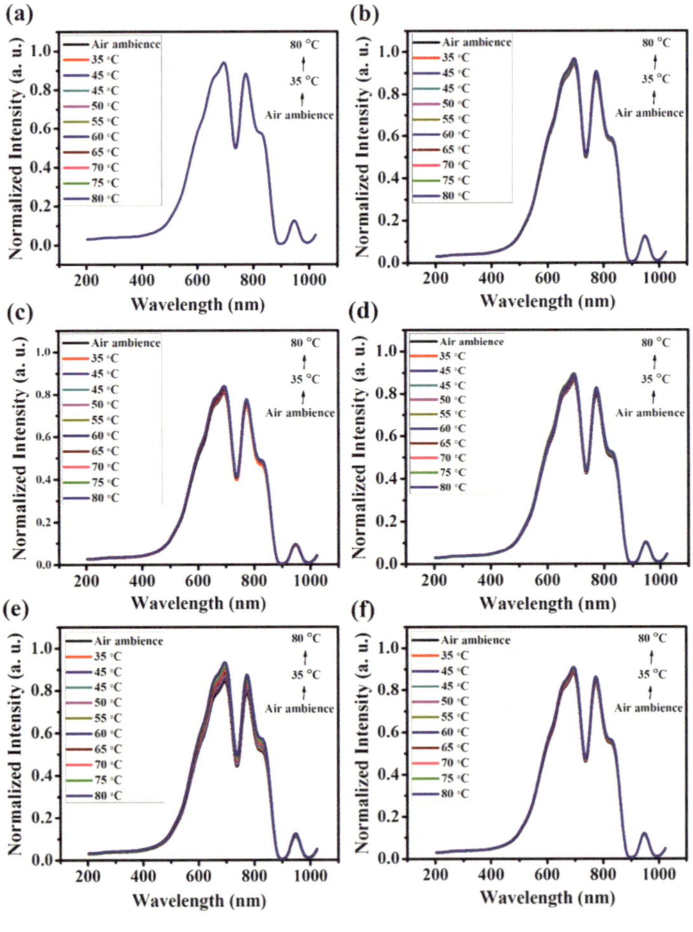

Figure 10.
Spectral response of the sensor at different temperature range of 35–80°C (a) bare fiber (b) Al_2O_3 (c) MgO (d) Al_2O_3-MgO (25–75%) (e) Al_2O_3-MgO (50–50%) and (f) Al_2O_3-MgO (75–25%). The arrow mentioned in the figure indicates the increase in intensity of the spectra upon increase in temperature [45].

DOI: http://dx.doi.org/10.5772/intechopen.110496

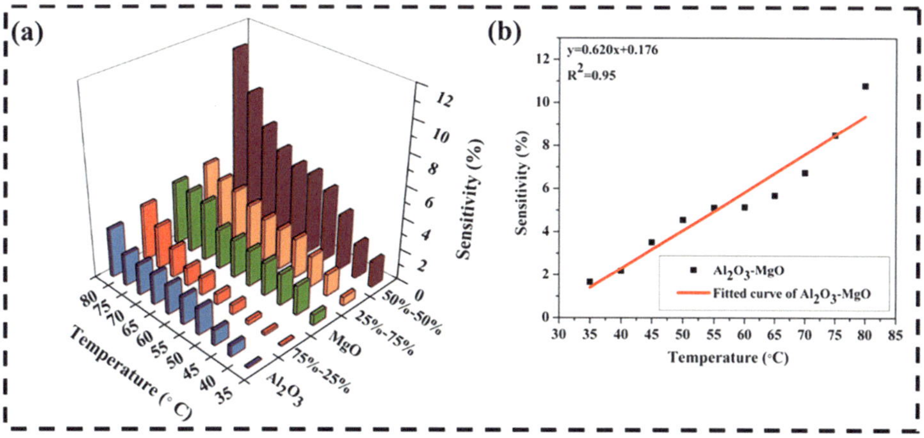

Figure 11.
(a) Temperature sensitivity of Al_2O_3, MgO, Al_2O_3-MgO (25–75%), Al_2O_3-MgO and Al_2O_3-MgO (75–25%) nanoparticles at the wavelength of 693 nm. (b) Relationship between temperature and sensitivity of Al_2O_3-MgO (50–50%) [45].

MgO nanoparticles are in rhombohedral structure and cubic structure and Al_2O_3-MgO nanocomposite conceives both rhombohedral and cubic crystal structure. The SEM morphology of Al_2O_3, MgO, and Al_2O_3-MgO (25–75%, 50–50%, 75–25%) composite nanopowders showed non-uniform agglomerated nanoparticles. The EDS analysis addresses the distribution of Al, Mg, and O elements, respectively in Al_2O_3-MgO nanocomposite. The unclad part of an optical fiber was coated with Al_2O_3, MgO, and Al_2O_3-MgO nanocomposites to create the temperature sensor probe. The temperature sensor response has been studied in the temperature range of 35–80°C (**Figure 10**) and Al_2O_3-MgO (50–50%) composite demonstrated an ultrahigh sensitivity of 0.62%/°C with a better linear regression coefficient of 95% (**Figure 11**). Further, the fabricated sensor emphasizes the feature of compact sensing structure, high-temperature sensitivity, good linearity, and wide temperature measurement range.

4. Challenges and future prospects

The future development of fiber optic temperature sensor faces challenges and opportunities, such as: (1) The development of new high-temperature-resistant optical fiber with excellent material and mechanical properties improving the temperature range of the sensor; (2) Adding durable cladding to crystal fiber highly enhances the sensing performance and long-term stability and (3) The development of sensors with multi-parameters sensing is essential along with better stability and package protection.

5. Conclusion

In summary, Al_2O_3-based nanomaterials have attracted significant attention of the scientific community for the fabrication of fiber optic-based temperature sensors due to their distinct electrical, mechanical, thermal, and optical properties. This article has discussed the crucial function of Al_2O_3-based nanomaterials in evanescent wave-based temperature sensors, as well as their present advancements and difficulties. Al_2O_3-based nanoparticles serve as a better sensing layer, sensitivity enhancement material, and sensing matrix material in fiber optic-based temperature sensors.

Acknowledgements

The authors would like to express their heartfelt gratitude to the Department of Science and Technology (DST), New Delhi, India, for providing financial assistance through the FIST (Fund for Improvement of S&T Infrastructure in Higher Education Institutions) project [SR/FST/ETI-015/2011].

Conflict of interest

The authors declare no conflict of interest.

Author details

Subramaniyam Narasimman[1], Lakshmi Narayanan Balakrishnan[2], Arunkumar Chandrasekhar[3] and Zachariah C. Alex[3]*

1 Department of Electronics and Communication Engineering, Sri Venkateswara College of Engineering and Technology (Autonomous), Chittoor, India

2 Department of Physics, Government College of Technology, Coimbatore, India

3 School of Electronics Engineering, VIT, Vellore, India

*Address all correspondence to: zcalex@gmail.com

References

[1] Dhanalakshmi S, Chakravartula V, Narayanamoorthi R, Kumar R, Dooly G, Duraibabu DB, et al. Thermal management of solar photovoltaic panels using a fibre Bragg grating sensor-based temperature monitoring. Case Studies in Thermal Engineering. 2022;**31**:101834

[2] Michalski JML, Eckersdorf K, Kucharski J. Temperature Measurement. 2nd ed. Chichester: Wiley; 2001

[3] Webb RC, Bonifas AP, Behnaz A, Zhang YH, Yu KJ, Cheng HY, et al. Ultrathin conformal devices for precise and continuous thermal characterization of human skin. Nature Materials. 2013;**12**:938-944

[4] Fang CF, Lee FD, Stober B, Fuller GG, Shen AQ. Integrated microfuidic platform for instantaneous flow and localized temperature control. RSC Advances. 2015;**5**:85620-85629

[5] Hill KO, Meltz G. Fiber Bragg grating technology fundamentals and overview. Journal of Lightwave Technology. 1997;**15**:1263-1276

[6] Mihailov SJ. Fiber Bragg grating sensors for harsh environments. Sensors. 2012;**12**:1898-1918

[7] Zhu T, Wu D, Liu M, Duan DW. In-line fiber optic interferometric sensors in single-mode fibers. Sensors. 2012;**12**:10430-10449

[8] Li E, Wang X, Zhang C. Fiber-optic temperature sensor based on interference of selective higher-order modes. Applied Physics Letters. 2006;**89**:091119

[9] Choi HY, Park KS, Park SJ, Paek UC, Lee BH, Choi ES. Miniature fiber-optic high temperature sensor based on a hybrid structured Fabry–Perot interferometer. Optics Letters. 2008;**33**:2455-2457

[10] Li M, Liu Y, Zhao X, Gao R, Li Y, Qu S. High sensitivity fiber acoustic sensor tip working at 1550 nm fabricated by two-photon polymerization technique. Sensors and Actuators A. 2017;**260**:29-34

[11] Liu G, Han M, Hou W. High-resolution and fast-response fiber-optic temperature sensor using silicon Fabry-Pérot cavity. Optics Express. 2015;**23**:7237-7247

[12] Zhu Y, Wang A. Surface-mount sapphire interferometric temperature sensor. Applied Optics. 2006;**45**:6071-6076

[13] Nguyen LV, Warren-Smith SC, Ebendorff-Heidepriem H, Monro TM. Interferometric high temperature sensor using suspended-core optical fibers. Optics Express. 2016;**24**:8967-8977

[14] Xu F, Brambilla G. Manufacture of 3-D microfiber coil resonators. IEEE Photonics Technology Letters. 2007;**19**:1481-1483

[15] Huang C, Lee D, Dai J, Xie W, Yang M. Fabrication of high-temperature temperature sensor based on dielectric multilayer film on sapphire fiber tip. Sensors and Actuators A. 2015;**232**:99-102

[16] Botewad SN, Pahurkar VG, Muley GG. Fabrication and evaluation of evanescent wave absorption based polyaniline-cladding modified fiber optic urea biosensor. Optical Fiber Technology. 2018;**40**:8-12

[17] Sun H, Hu M, Rong QZ, Du Y, Yan H, Qiao X. High sensitivity optical

fiber temperature sensor based on the temperature cross-sensitivity feature of RI-sensitive device. Optics Communication. 2014;**323**:28-31

[18] Rahman A, Panchal K, Kumar S. Optical sensor for temperature measurement using bimetallic concept. Optical Fiber Technology. 2011;**17**:315-320

[19] Zhu L, Zeng W. Room-temperature gas sensing of ZnO-based gas sensor: A review. Sensors and Actuators A. 2017;**267**:242-261

[20] Tong X, Shen W, Chen X, Corriou J. A fast response and recovery H2S gas sensor based on free-standing TiO2 nanotube array films prepared by one-step anodization method. Ceramics International. 2017;**43**:14200-14209

[21] Kou X, Xie N, Chen F, Wang T, Guo L, Wang C, et al. Lu G superior acetone gas sensor based on electrospun SnO2 nanofibers by Rh doping. Sensors and Actuators B: Chemical. 2018;**256**:861-869

[22] Guo X, Pan G, Ma X, Li X, He P, Hu Z, et al. Optical excitation-enhanced sensing properties of acetone gas sensors based on Al2O3-doped ZnO. Sensor Review. 2017;**37**:364-370

[23] Rao CN, Nakate UT, Choudhary RJ, Kale SN. Defect-induced magneto-optic properties of MgO nanoparticles realized as optical fiber-based low-field magnetic sensor. Applied Physics Letters. 2013;**103**:151107

[24] Bai Z, Chen R, Si P, Huang Y, Sun H, Kim D. Fluorescent pH sensor based on Ag@SiO2 core–shell nanoparticle. ACS Applied Materials & Interfaces. 2013;**5**:5856-5860

[25] Ye F, Tian C, Ma C, Zhang ZF. Fiber optic sensors based on circular and elliptical polymer optical Fiber for measuring refractive index of liquids. Optical Fiber Technology. 2022;**68**:102812

[26] Chen CH, Tsao TC, Tang JL, Wu WT. A multi-D-shaped optical fiber for refractive index sensing. Sensors. 2010;**10**:4794-4804

[27] Szczerska M. Temperature sensors based on polymer fiber optic interferometer. Chem. 2022;**10**:228

[28] Prasad SG, Lal C. Spectroscopic investigations of optical bandgap and search for reaction mechanism chemistry due to-rays irradiated PMMA polymer. Biointerface Research in Applied Chemistry. 2022;**13**:184

[29] Zuo H, Yu S, Gu T, Hu J. Low loss, flexible single-mode polymer photonics. Optics Express. 2019;**27**:11152-11159

[30] Gutierrez-Rivera M, Jauregui-Vazquez D, Garcia-Mina DF, Sierra-Hernandez JM, EstudilloAyala JM, Almanee M, et al. Fiber optic fabry-perot micro-displacement sensor based on low-cost polymer film. IEEE Sensors Journal. 2020;**20**:4719-4725. DOI: 10.1109/JSEN.2019.2944998

[31] Chapalo I, Theodosiou A, Kalli K, Kotov O. Multimode Fiber interferometer based on graded-index polymer CYTOP fiber. Journal of Lightwave Technology. 2020;**38**:1439-1445

[32] Gierej A, Geernaert T, Van Vlierberghe S, Dubruel P, Thienpont H, Berghmans F. Challenges in the fabrication of biodegradable and implantable optical fibers for biomedical applications. Materials. 2021;**14**:1972

[33] Woyessa G, Fasano A, Stefani A, Markos C, Nielsen K, Rasmussen HK, et al. Single mode step-index polymer

optical Fiber for humidity insensitive high temperature Fiber Bragg grating sensors. Optics Express. 2016;**24**:1253-1260

[34] Min R, Hu X, Pereira L, Simone Soares M, Silva LCB, Wang G, et al. Polymer optical Fiber for monitoring human physiological and body function: A comprehensive review on mechanisms, materials, and applications. Optics and Laser Technology. 2022;**147**:107626

[35] Jedrzejewska-Szczerska M. Response of a new low-coherence Fabry-Perot sensor to Hematocrit levels in human blood. Sensors. 2014;**14**:6965-6976

[36] Mizuno Y, Theodosiou A, Kalli K, Liehr S, Lee H, Nakamura K. Distributed polymer optical fiber sensors: A review and outlook. Photonics Research. 2021;**9**:1719

[37] Gomez D, Morgan SP, Hayes-Gill BR, Correia RG, Korposh S. Polymeric optical fibre sensor coated by SiO_2 nanoparticles for humidity sensing in the skin microenvironment Sens. Actuators B. 2018;**254**:887-895

[38] Zhong N, Wang Z, Chen M, Xin X, Wu R, Cen Y, et al. Three-layer-structure polymer optical fiber with a rough inter-layer surface as a highly sensitive evanescent wave sensor Sens. Actuators B. 2018;**254**:133-142

[39] Huang YW, Tao J, Huang XG. Research Progress on F-P interference-based Fiber-optic sensors. Sensors. 2016;**16**(9):1424

[40] Villalba A, Martín JC. Interferometric temperature sensor based on a water-filled suspended core fiber. Optical Fiber Technology. 2017;**33**:36-38

[41] Zeng X, Wu Y, Hou C, Bai J, Yang G. A temperature sensor based on optical microfiber knot resonator. Optics Communication. 2009;**282**:3817-3819

[42] Barakat WS, Wagih A, Elkady OA, Abu-Oqail A, Fathy A, EL-Nikhaily A. Effect of Al_2O_3 nanoparticles content and compaction temperature on properties of Al–Al_2O_3 coated Cu nanocomposites. Composites Part B: Engineering. 2019;**175**:107140

[43] Taneja S, Punia P, Thakur P, Thakur A. Synthesis of nanomaterials by chemical route. In: Thakur A, Thakur P, Khurana SP, editors. Synthesis and Applications of Nanoparticles. Singapore: Springer. DOI: 10.1007/978-981-16-6819-7_4

[44] Krishnia L, Thakur P, Thakur A. Synthesis of nanoparticles by physical route. In: Thakur A, Thakur P, Khurana SP. Synthesis and Applications of Nanoparticles. Singapore: Springer. DOI: 10.1007/978-981-16-6819-7_3

[45] Narasimman S, Balakrishnan L, Alex ZC. Clad-modified fiber optic ammonia sensor based on Cu functionalized ZnO nanoflakes. Sensors and Actuators A: Physical. 2020;**316**:112374

[46] Narasimman S, Harish Babu K, Balakrishnan L, Meher S. R, Sivacoumar R, Alex, ZC. Fabrication of fiber optic based temperature sensor. Materials Today: Proceedings. 2019;9:164-174

[47] Narasimman S, Mahararana K, Kokila SK, Balakrishnan L, Alex ZC. Al_2O_3-MgO nanocomposite based fiber optic temperature sensor. Materials Research Express. 2018;**5**:115014

Chapter 3

Plant Base Renewable Energy to Power Nanoscale Sensors

Ajay Kumar Singh

Abstract

The modern technologies have been revolutionized due to tremendous progress in Internet-of-Things (IoT). Sensors are a core component to make a bridge between the Internet and surrounding environments. The progress in power efficient communication network makes it possible to deploy the sensors in remote areas. The major drawback of these sensors is that they use Li-ion battery for power supply, which needs frequent recharging/replacement due to massive number of connected devices to IoT. The hazardous chemicals left in environment after the use of battery is another concern. Since modern nanoscale sensors need only nanoscale power (of order of μWatt), nanogenerators can play an important role to provide self-powered sensors, which is growing technology that can harvest small-scale energy from piezo- and pyroelectric effect. However, this technique is lightweight but not cost-effective and biodegradable. We have proposed a green, sustainable energy harvesting system based on living plants because plants are the undisputed champion of solar power that operates at nearly 100% efficiency. Plant-based energy generation is a method that harvests electrical energy from living plants due to a chemical reaction between the plant and a pair of electrodes. This energy is available 24×7 day and night irrespective of environmental conditions.

Keywords: renewable and sustainable energy, living plants, *Sansevieria trifasciata* plant, *Aloe vera* plant, *Beschorneria* plant, sensor nodes, plant base cell

1. Introduction

The Internet makes the world a small village that permits the interaction between smart things and the effective integration of real-world information and knowledge in the digital world. These things include not only communication devices, but also physical objects, like cars, computer, and home appliances, which are controlled through wireless communication networks (WSNs). The Internet of Things (or IoT) refers to the connection of billions of physical devices such as sensors, actuators, mobile phones, drones, etc., to the internet to collect and share the data for making collaborative decisions to accomplish the tasks in an optimal manner [1, 2]. Due to the growing awareness of environmental issues around the world, green IoT technology initiatives should be taken into consideration. Green IoT focuses on reducing IoT energy usage which is a necessity for fulfilling the requirement of reducing CO_2

emissions. Harvesting energy from non-conventional sources in the environment has received attention over the past decade due to miniaturization of the devices which makes it possible to consume lower power (of order of mW/nanoW) in many applications. Traditionally, sensors, wearable, and portable electronic devices, mobile phones, automatic security systems, etc., need Li-ion battery for their external power supply which is not perpetual and free of maintenance because the battery has a limited lifetime and needs frequent recharging and replacement. Researchers have proposed nanogenerators for harvesting energy from the ambient mechanical motion to make the sensors self-powered [3–6]. Nanogenerators are generally an energy-harvesting device that generate electricity from waste mechanical energy [7–9]. There are several ambient energy-harvesting techniques based on the piezo-electric effect, triboelectric effect, pyroelectric effect, and electromagnetic induction [10–13]. Nanogenerators are an evolving energy-harvesting technology that convert various forms of mechanical energy such as human motion, vibration, flowing water, raindrops, wind, and waste heat into electrical energy [14–16]. This technique is not cost-effective as well as not available 24×7 days. Due to these reasons, it is necessary to look at some other alternative green sources for autonomous self-powered sensors. Although, harvesting electrical energy from the sun is a matured and well-accepted technique, [17–19] this technique has a limitation that it remains functional only in the presence of sunlight. Recently scientific explorations showed that plants may become a potential source of bioenergy that is not only renewable but also sustainable and cheap [20–24]. Plants are called autotrophs because they can use energy from light to make their own food. In the presence of light and chlorophyll, water, and carbon di-oxide (CO_2) are chemically combined in leaf of plants to make glucose. The produced glucose supports the growth of the plants. This process is called photo-synthesis and is performed by all plants [25–27]. Respiration in plants, on the other hand, is a reverse process of photosynthesis in which glucose molecules (obtained during photosynthesis process) are broken down in presence of oxygen to liberate energy [28–30]. These two processes induce the flow of electrons inside the plants which can be captured by pair of electrodes to harvest electrical potential [31–33]. By embedding electrodes into the plants and employing an electrochemistry process, the chemical energy can be converted into electrical energy via an oxidization-reduction reaction [34] process. The oxidization process, which happens at the anode electrode, and reduction process, which happens at the cathode electrode, causes the electron to flow from anode to cathode to produce electricity. This system is termed plant-based cell (PBC) which provides a direct method to harvest DC electrical energy from living plants which can be potentially used to power up ultra-low-power devices and IoT sensor nodes in the future. The rate of photosynthesis and respiration processes are influenced by external environmental factors, such as concentration of oxygen and carbon-di-oxide in the air, amount of water in the soil as well as nutrient conditions of soil [35, 36]. Other external stimuli, like stress due to wound, temperature, light intensity variations, and water disparity also influence the harvested electrical potential in the plant. The succulent plant produces much higher voltage compared to non-succulent plant because CAM (crassulacean acid metabolism) plants contain more rubisco genes (or chloroplasts) [37, 38]. Succulent plants are water-retaining plants, which can store water in their leaves, stems, and roots to survive in a dry environment. *Aloe barbadensis* Miller (*Aloe Vera*), *Beschorneia* and *Sansevieria* plants belong to succulent family [39, 40]. These plants can grow in tropical, sub-tropical, warmer temperature regions, and exchange the oxygen and CO_2 using CAM process at night. The CAM process allows them to withstand drought because their stomata open only at night to prevent the water from escaping via evaporation in the hot sun [41]. Researchers have paid attention to harvesting electrical energy from *Aloe vera* plants [42–44]. The major disadvantages of *Aloe vera* plants are that once the electrodes are inserted in leaf, leaf dyes faster and its survival rate is very short due to chemical reaction between electrode and the gel of *Aloe vera* despite its hardness and unique self-repairing ability. It also does not grow in wild but needs to be cultivated. Due to these reasons, attention has been given to *Sansevieria trifasciata* and *Beschorneia* plants to harvest electrical energy. Most *Sansevierias* are native to Africa, although a few originated in India and Asia. This plant is a very aggressive invasion plant and is able to grow in a great range of sunlight to a partially shaded area. The leaf of *Sansevieria trifasciata* plant can survive longer and even self-repaired if any wound happens after inserting the electrodes [45]. *Beschorneria* plant is a stemless plant with 20–35 linear, lanceolate, leathery leaves that are widened at their base. They are gray-green to green, about 40–60 cm long. The leaf margins are finely denticulate [46]. These two plants have been ignored while harvesting electrical energy compared to *Aloe vera* plant.

In this chapter, first time we have performed an experiment on *Sansevieria* and *Beschorneria* plants to harvest electrical energy to power up the IoT sensor nodes. A detailed study was carried out to see the effect of external and internal environmental factors on the harvested potential. The *Beschorneria* plant alone results in larger potential (0.96 V) in dry soil whereas *Sansevieria* 3 plant gives larger potential (V = 1.05 V) in presence of acidic soil which is a favorable condition for plant's growth. The harvested potential reduces sharply under stress conditions like wet soil, wound leaf, etc. The harvested potential is 2.70 V when three different succulent plants were connected in series. By taking a suitable combination (parallel/or series) of *Sansevieria* and *Beschorneria* plants, appropriate electrical energy can be harvested to power up the IoT sensor node.

2. Experimental setup

Figure 1a shows the schematic representation whereas **Figure 1b** illustrates the actual experimental setup to study the various aspects that influence the harvested energy from the succulent plants. We have chosen *Sansevieria trifasciata*, commonly called **snake plant or mother-in-law's tongue, and** *Beschorneria* plant to conduct the experiment because these two succulent plants provide larger electrical energy compared to *Aloe vera* plant. The average leaf size of these two plants varies from 30 to 90 cm in length. All the experiments have been performed in an indoor laboratory unless and until specified with room temperature of 28–30°C and average humidity of 49%. The plants were located closer to a closed transparent glass window to adjust the light intensity. We have used aluminum (Al) and copper (Cu) as an electrode pair in form of a sheet. These electrode pairs are regularly cleaned to remove contaminants if any. The harvested potential is measured under various conditions. First, electrical potential was harvested from single leaf of *Sansevieria* plant and *Beschorneria* plant, but the harvested current was very low. In the second scenario, we have used two/or three succulent plants in series, parallel or series-parallel combinations. We have chosen *Aloe vera*, Banana, and Cactus plants for comparison. Due to the close relationship between environmental factors and the electrical signal in plants, we have performed various experiments to monitor the effect of changes in the external environment on the harvested electrical potential.

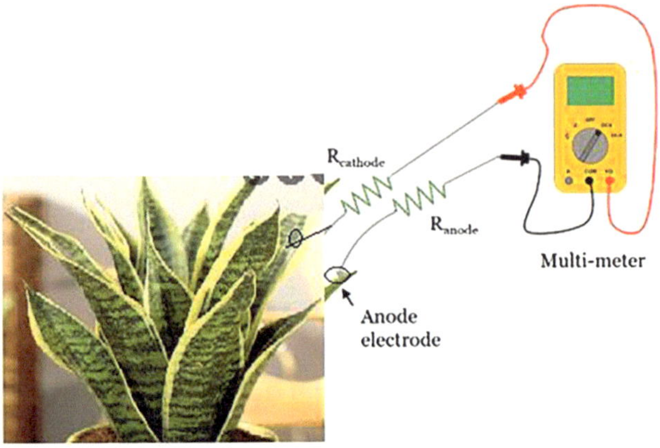

Figure 1.
Experimental setup (a) Schematic representation (b) Actual setup.

3. Results and discussion

The shape of the electrode is an important factor to decide the electrical potential harvested from the living plants. During the experiment, we observed that the nail shape and sheet shape electrode pairs produce the same amount of potential (≥0.92 V) whereas touch electrode results in lower harvested potential (of order of 10 mV). Due to this reason, we have chosen a sheet shape of electrodes to conduct the experiment.

We have conducted the experiment on the soft and hard leaves of the *Sansevieria trifasciata* plant. It was observed that the maximum potential (V = 0.92 V) is obtained when one electrode is near root and other is on the edge of green/soft leaf compared to matured leaves (V = 0.88 V). This is due to the fact that in soft leaf the tip remains green and fleshy while in matured/or hard leaf tip becomes brown and woody which produces less glucose and results in lower number of electrons.

In general, the harvested potential from plant does not remain stable but varies with time as seen in **Figure 2a**. The larger and more stable potential is observed when both electrodes were inserted near the edge of two different soft leaves of *Sansevieria trifasciata* plant due to the constant photosynthesis process near the edge.

Figure 2b shows the effect of soil condition on the harvested potential. The result concludes that larger electrical potential ($V_{harvested}$ = 1.05 V) is obtained from a single *Sansevieria* leaf when soil is acidic in nature because the acidic soil provides favorable conditions for plant's growth [41]. This potential reduces to 0.9 V when excessive water is poured into the soil due to stress conditions.

During our experiment on various succulent plants, we have observed that the position of electrode pair on leaf affects the harvested potential. The *Aloe vera* plant produces 0.76 V from a single leaf which increases to 0.90 V when two electrodes are far away from each other whereas the cactus produces only 0.88 V for the same condition. We observed 0.92 V in case of *Sansevieria* plant when one electrode is near root and the other is near edge of a single green leaf whereas the *Beschorneria* plant results in 0.96 V for the same condition.

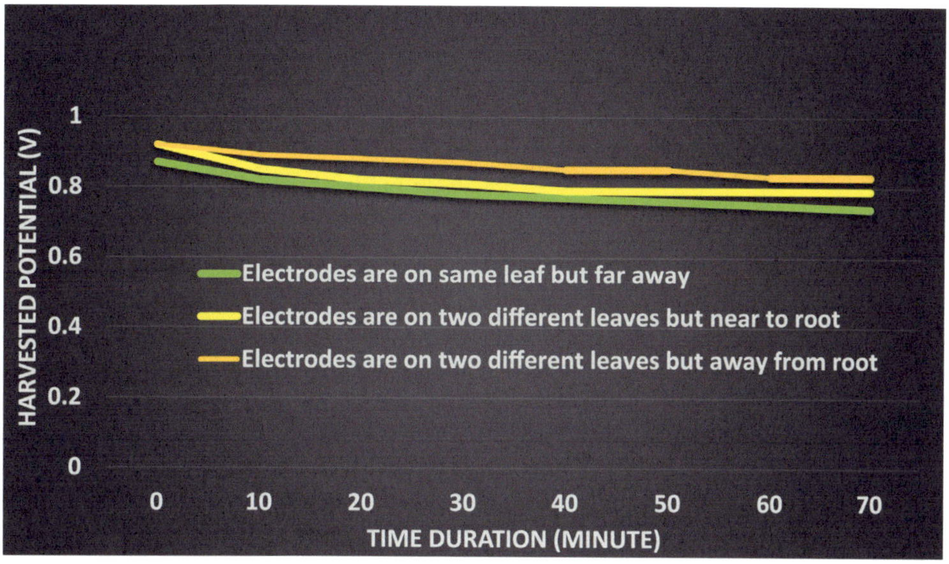

(a)

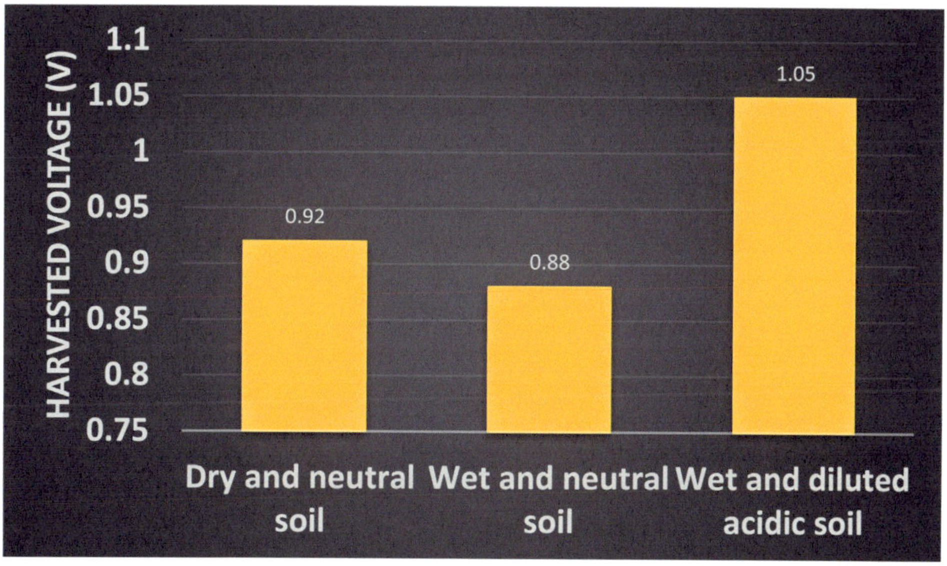

(b)

Figure 2.
(a) Variation of harvested potential with time. (b) Harvested potential from different soil conditions.

In another experiment, we connected two *Sansevieria trifasciata* plants in series by choosing the soft and matured leaf combination. The experiment results in 1.78 V and 38 μA current. The current increases to 45 μA when *Sansevieria* and *Beschorneria* plants were connected in series on the cost of reduced harvested voltage ($V_{harveste}$ = 1.70 V). The harvested electrical potential reaches 2.44 V when three *Sansevieria trifasciata* plants were connected in series, but the current was only 22 μA. The maximum electrical potential of 2.75 V was observed when *Sansevieria trifasciata*,

Aloe vera, and *Beschorneria* plants were connected in series. This is due to different photosynthetic rates. Therefore, it is preferable to connect different plants in series to harvest maximum electrical energy instead of using only one type of plant. The harvested potential falls drastically if Cu electrode is inserted deep inside the soil due to reduced bacterial activity.

Since *Sansevieria trifasciata* and *Beschorneria* plants were favorable succulent plants to harvest larger electrical potential, hence we have taken five possible combinations of these two plants as shown in **Figure 3**. Here, plant is represented by a cell.

Table 1 gives the experimental results of harvested potential and current in these five cases. The optimum condition for current and voltage is in case 4 which results in 72 μW electrical power due to the release of a larger number of electrons in presence of a higher photosynthesis rate.

Table 2 gives the experimental results of three series of connected succulent plants (two *Sansevieria* plants and one *Aloe vera* plant) in the presence of stress. Here, stress reflects the amount of damage or wound in the leaf after inserting the electrodes. The results show that the damage in the leaf reduces the harvested potential drastically because when all three series-connected leaves are damaged, the harvested potential is only 320 mV compared to 2.50 V when only *Aloe vera*'s leaf is damaged. Since the self-repair ability of *Sansevieria* and *Beschorneria* plants is larger than *Aloe vera* plant, therefore it is preferable to connect these two plants either in series or parallel rather than *Sansevieria*/*Beschorneria* and *Aloe vera* plants.

The size of *Beschorneria* leaf decides the harvested potential as seen in **Figure 4**. The harvested potential increases for lower leaf width (≤4 cm) and takes a lower value for width larger than 5 cm. This is due to a change in the internal resistance of the leaf. The harvested potential remains constant for 4 cm < W ≤ 5 cm. The whole experiment was performed by inserting an Al electrode on leaf and a Cu electrode inside the dry soil.

Figure 5 shows the variation of harvested power from succulent plants against the load resistance for ten different combinations (**Table 3**: case means series). The experimental results predict that maximum power can be harvested at R_L = 10 KΩ

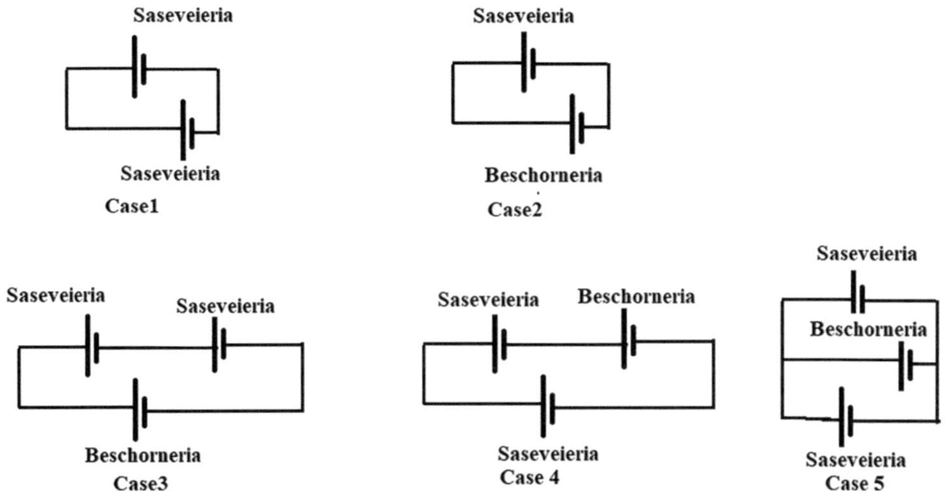

Figure 3.
Different combinations of Sansevieria and Beschorneria plants.

Case	Harvested potential (V)	Current (μA)
1	0.82	70
2	0.72	60
3	1.62	23
4	1.0	72
5	0.81	65

Table 1.
Harvested potential and current for four cases.

SN	Condition of leaf			Harvested potential (V)
	First *Sansevieria* plant	Second *Sansevieria* plant	*Aloe vera* plant	
1	Wounded	Slightly wounded	Fresh	0.41
2	Wounded	Slightly wounded	Wounded	0.35
3	Wounded	Wounded	Wounded	0.32
4	Fresh	Wounded	Wounded	2.08
5	Fresh	Fresh	Wounded	2.50
6	Fresh	Wounded	Fresh	1.95

Table 2.
Harvested electrical potential from three series-connected plants under stress conditions.

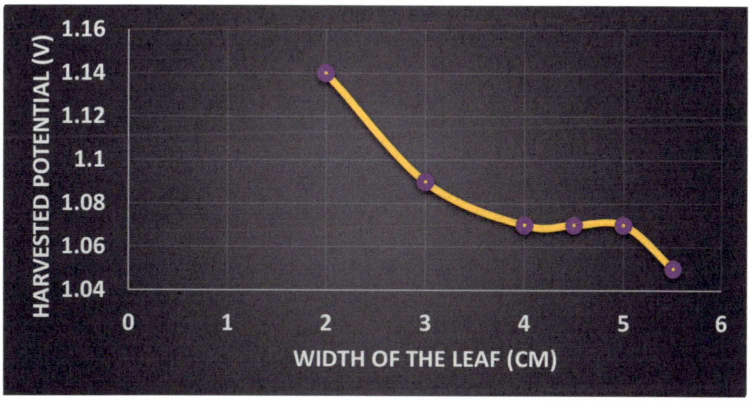

Figure 4.
Harvested potential versus width of leaf.

for series 6 which is about 72 μW whereas lower energy results for series 7. In the case of series 4, we have observed maximum variation in harvested energy at lower load resistance. These results only reflect that by considering different combinations of *Sansevieria* and *Beschorneria* plants for different loads we can harvest sufficient electrical energy to power up the sensor node. The average short circuit current is only of the order of 5 μA.

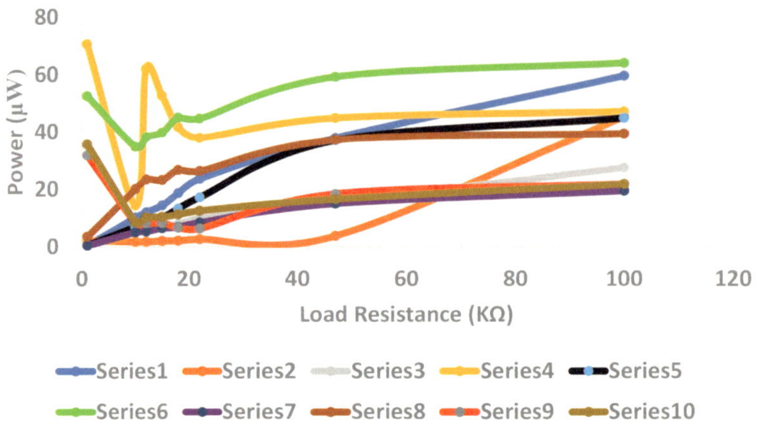

Figure 5.
Harvested electrical power against load for various combinations.

Case	Condition
1	Two *Sansevieria* plants are in series
2	One *Sansevieria* plant and one *Beschorneria* plant are in series
3	Two *Sansevieria* plant and one *Beschorneria* plants are in series
4	Two *Sansevieria* plants are in parallel
5	One *Sansevieria* plant and one *Beschorneria* plant are in parallel
6	One *Sansevieria* plant and one *Beschorneria* plant are in series and then parallel to another *Sansevieria* plant
7	Two *Sansevieria* plants are in series and then parallel to *Beschorneria* plant
8	Two *Sansevieria* plants and one *Beschorneria* plant are in parallel
9	Single *Sansevieria* plant
10	Single *Beschorneria* plant

Table 3.
Different cases for harvesting electrical power.

The voltage-current characteristics is shown in **Figure 6** for different cases. Here series 1 means two *Sansevieria* plants are in series, series 2 indicates one *Sansevieria* plant and one *Beschorneria* plant in series whereas case 3 reflects two *Sansevieria* plants in series and then in parallel with *Beschorneria* plant. Lastly, series 4 represents case 6 as discussed in **Table 3**. From the results, we observed that irrespective of the harvested voltage, current is always larger in case 4 which also results in larger electrical energy.

We have charged the capacitor at 100 μF using the experimental setup as described for series 4 in **Figure 3**. **Figure 7** shows the voltage and current at various time intervals. The minimum current and voltage results in the next day at about 12 noon due to the heat effect which lowers the photosynthesis activity.

For the estimation of oxygen releasing potential on the harvested electrical energy from *Sansevieria* plant, we have kept the plant inside the leakproof airtight green polythene bag at ambient temperature to create light conditions. The

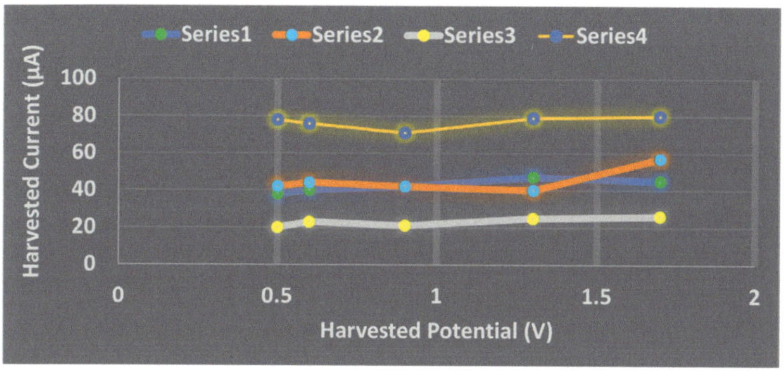

Figure 6.
Harvested current versus harvested voltage for different combination.

electrodes were inserted suitably to avoid any leakage of gas. After one day, we measured the average harvested potential of about 0.89 V. Dark condition is created by covering the plant with black bag along with green polythene bag. Mouth of the polythene bag was tied tightly with a provision to measure the harvested potential using multimeter in both cases. After three days, we have measured about 0.84 V in this condition whereas after 6 days the harvested potential drops to 280 mV. This shows that the *Sansevieria trifasciata* plant maintains its oxygen level even in the dark condition for a limited period and the harvested electrical potential is not affected in dark light for that period.

Light intensity is the main factor that control the central process of plant such as photosynthesis. To see the effect of light intensity on the harvested potential, we have put the *Sansevieria trifasciata* in a chamber and illuminated it with 15 W incandescent yellow and red bulbs separately. We also illuminated the plant using a 7 W LED bulb. The whole experiment was performed for 5 hours. We have measured the relative percentage change in the harvested potential after every 30 minutes using formula given by equation (1).

$$\%\,variation\ in\ harvested\ potential = \left(\frac{final harvested\ potential - intial\ harvetsed\ potential}{initial\ harvested\ potential}\right) \times 100 \quad (1)$$

As seen from **Figure 8**, the harvested potential decreases as the exposure time increases except when plant is exposed to 15 W incandescent yellow bulb. In the case of 15 W yellow bulb, the harvested potential initially increases up to 2 hours due to increased energy of associated photon and then falls gradually due to excessive heat which reduces the photosynthesis activity in the plant. The harvested electrical potential shows a sharp decrease with exposure time in the case of 15 W red bulb due to the reduced energy of associated photon.

We have also conducted the experiment in outdoor conditions at 3 PM in summer for following cases; when three *Sansevieria* plants are connected in series the harvested potential is 2.66 V which is larger than 2.46 V in the lab condition due to increased photosynthesis activity whereas when one *Sansevieria*, one *Aloe vera*, and one *Beschorneria* plants are connected in series, the harvested potential is 2.73 V which is lower than 2.77 V as in lab condition.

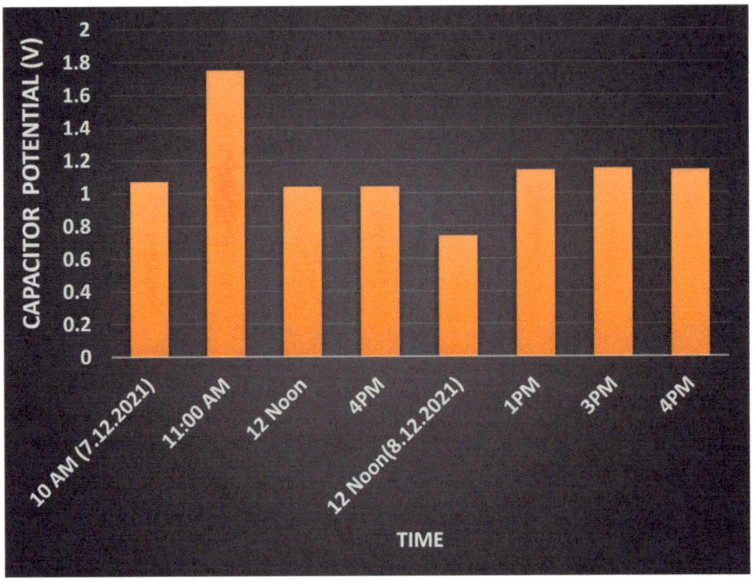

(a)

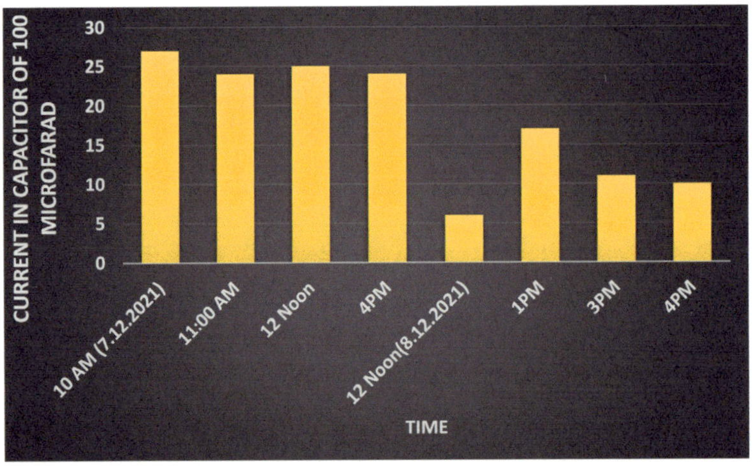

(b)

Figure 7.
Charging of capacitor at different time intervals. (a) Harvested potential in capacitor at different time interval. (b) Charging of capacitor 100 μF.

4. Conclusion

Succulent plants are favorable candidates to replace Li-ion battery in future to power up embedded IoT sensors. The chosen *Sansevieria* and *Beschorneria* plants for harvesting electrical potential is due to their higher self-repairing capability and photosynthesis rate compared to *Aloe vera* plant. The combination of these two plants harvests a larger potential than other succulent plants due to higher conductance of CO_2 for photosynthesis. The harvested potential is more than 1 V from a single leaf of *Sansevieria* plant in presence of dilute acidic soil which is favorable condition for plant's growth. The experimental findings suggest that the series and parallel

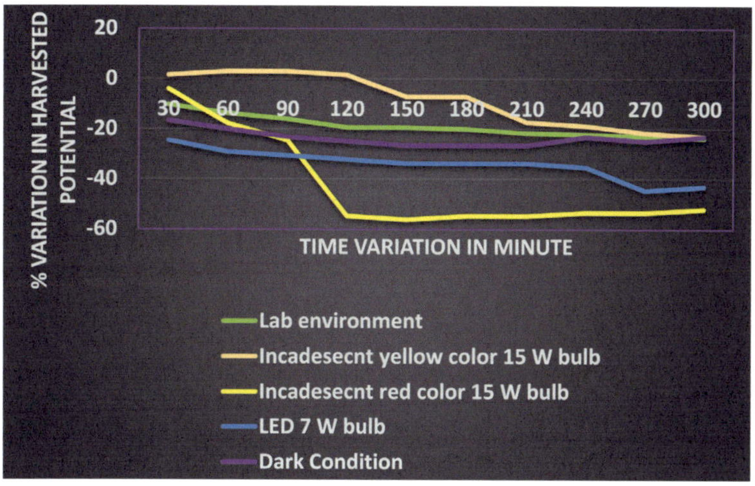

Figure 8.
Percentage variation of harvested potential with illumination exposure time.

combination of the succulent plants affect the harvested energy along with external stimuli, nature of soil, number of connected leaves, and position of electrodes. We have observed that if different types of succulent plants are connected either in series or parallel compared to same type of plants, the harvested energy is more. Maximum power we harvested during our experiment was 72 μW which is sufficient to power up the embedded sensors which can be further enhanced by optimizing the various factors affecting the harvested energy. The harvested electrical potential falls with time as well as the with the separation between two electrodes due to associated parasitic capacitance and resistance. The wound in the leaf affects the harvested potential severely. The CO_2 content and intensity of light are other factors that affect the harvested potential from succulent plants. This study confirms that succulent plants like *Sansevieria* and *Beschorneria* plants prove themselves as the future candidate for green energy to replace the conventional battery to power up sensor nodes.

Acknowledgements

Author is thankful to Dr. Narayan Kumar, Biotechnology and Bioinformatics Engineering-NIIT University for his valuable suggestion and discussions related to the plants. Author is also thankful to Mr. Narendra Singh Bisht (Lab in-charge) for providing all the help needed to conduct the experiment.

Author details

Ajay Kumar Singh
Electronics and Communication Engineering, NIIT University, Alwar, Rajasthan, India

*Address all correspondence to: aks_1993@yahoo.co.uk

References

[1] Kumar S, Tiwari P, Zymbler M. Internet of Things is a revolutionary approach for future technology enhancement: A review. Journal of Big Data. 2019;**6**(111):1-21

[2] Nižetić S, Šolić P, López-de-Ipiña González-de-Artaza D, Patrono L. Internet of Things (IoT): Opportunities, issues and challenges towards a smart and sustainable future. Journal of Cleaner Production. 2020;**274**:1-32

[3] Lin Z. Nanogenerators for Self-Powered Devices and Systems. Atlanta, USA: Georgia Institute of Technology; 2011

[4] Zi Y, Wang ZL. Nanogenerators: An emerging technology towards nanoenergy. APL Materials. 2017;**5**:074103

[5] Sripadmanabhan Indira S, Aravind Vaithilingam C, Oruganti KSP, Mohd F, Rahman S. Nanogenerators as a sustainable power source: State of art, applications, and challenges. Nanomaterials. 2019;**9**:1-35

[6] Adhikari A, Sengupta J. Nano tools and devices enhanced renewable energy. USA: Elseiver Publisher; 2021. ISBN: 9780128216996. DOI: 10.1016/B978-0-12-821709-2.00004-9

[7] Interesting Engineers. Nanogenerators: The Cleverest Things You've Never Seen. 2016. Available from: https://interestingengineering.com/nanogenerators-the-cleverest-things-youve-never-seen

[8] Lia M, Porterb AL, Wang ZL. Evolutionary trend analysis of nanogenerator research based on a novel perspective of phased bibliographic coupling. Nano. 2017;**34**:93-102

[9] Linda Crampton. Nanogenerator Facts and Charging Electronic Devices by Movement. 2021. Available from: https://turbofuture.com/industrial/Using-Muscle-Action-to-Power-Cell-Phones-iPods-and-Similar-Devices

[10] Briscoe J, Dunn S. Piezoelectric nanogenerators—A review of nanostructured piezoelectric energy harvesters. Nano Energy. 2015;**14**:15-29

[11] Vivekananthan V, Alluri NR, Chandrasekhar A, Purusothaman Y, Gupta A, Kim S-J. Zero-power consuming intruder identification system by enhanced piezoelectricity of $K_{0.5}Na_{0.5}NbO_3$ using substitutional doping of BTO NPs. Journal of Material Chemistry C. 2019;**7**:7563-7571. DOI: 10.1039/C8TC06626D

[12] Kim W-G, Kim D-W, Tcho I-W, Kim J-K, Kim M-S, Choi Y-K. Triboelectric nanogenerator: Structure, mechanism, and applications. ACS Nano. 2021;**15**(1):258-287

[13] Long L, Liu W, Wang Z, et al. High performance floating self-excited sliding triboelectric nanogenerator for micro mechanical energy harvesting. Nature Communications. 2021;**12**:4689

[14] Vivekananthan V, Kim WJ, Alluri NR, et al. A sliding mode contact electrification based triboelectric-electromagnetic hybrid generator for small-scale biomechanical energy harvesting. Micro and Nano System Letters. 2019;**7**(14):1-8. DOI: 10.1186/s40486-019-0093-6

[15] Kim WJ, Vivekananthan V, Khandelwal G, Chandrasekhar A, Kim S-J. Encapsulated triboelectric–electromagnetic hybrid generator for a sustainable blue energy harvesting

and self-powered oil spill detection. ACS Applied Electronic Materials. 2020;**2**(10):3100-3108

[16] Hajra S, Vivekananthan V, Sahu M, Khandelwal G, Raj NPMJ, Kim S-J. Triboelectric nanogenerator using multiferroic materials: An approach for energy harvesting and self-powered magnetic field detection. Nano Energy. 2021;**85**:105964

[17] Shaikh MR, Shaikh S, Waghmare S, Labade S, Tekale A. A review paper on electricity generation from solar energy. International Journal for Research in Applied Science and Engineering Technology. 2017;**5**:1884-1889

[18] Novas N, Garcia RM, Camacho JM, Alcayde A. Advances in solar energy towards efficient and sustainable energy. Sustainability. 2021;**13**:1-31

[19] Li D, King M, Dooner M, Guo S, Wang J. Study on the cleaning and cooling of solar photovoltaic panels using compressed airflow. Solar Energy. 2021;**221**:433-444

[20] Timmers RA. Electricity Generation by Living Plants in a Plant Microbial Fuel Cell. Wageningen: Wageningen University; 2012

[21] Gurram SPG, Kothapalli NS. A novel electricity generation with green technology by Plant-e from living plants and bacteria: A natural solar power from living power plant. In: 2017 6th International Conference on Computer Applications in Electrical Engineering-Recent Advances (CERA). Roorkee, India: IEEE; 5-7 October 2017. pp. 146-151

[22] Meder, Must I, Sadeghi A, Mondini A, Filippeschi C, Beccai L, et al. Energy conversion at the cuticle of living plants. Advanced Functional Materials. 2018;**28**:1-8

[23] Available from: https://techxplore.com/news/2015-01-plant-e-street-alive.html

[24] Muladi M, Jalil MFA, Arifin RF, Aripriharta A, Zaini IAE, Sendari S, et al. An experimental study of generating electricity from urban tropical forest plants. Journal of Physics: Conference Series. 2021;**1825**:1-7

[25] Flexer V, Manao N. From dynamic measurements of photosynthesis in living plant to sunlight transformation into electricity. Analytical Chemistry. 2010;**82**(4):1444-1449

[26] Kohli A, Miro B, Balié J, Hughes J. Photosynthesis research: A model to bridge fundamental science, translational products, and socio-economic considerations in agriculture. Journal of Experimental Botany. 2020;**71**(7):2281-2298

[27] Yavari N, Tripathi R, Wu BS, MacPherson S, Singh J, et al. The effect of light quality on plant physiology, photosynthetic, and stress response in *Arabidopsis thaliana* leaves. PLoS One. 2021;**16**(3):e0247380

[28] Gonzalez-Meler MA, Taneva L, Trueman RJ. Plant respiration and elevated atmospheric CO_2 concentration: Cellular responses and global significance. Annals of Botany. 2004;**94**(5):647-656

[29] Brendan M, Asao S, Millar AH, Atkin OK. Core principles which explain variation in respiration across biological scales. New Phytologist. 2019:670-686

[30] Tcherkez G, Atkin OK. Unravelling mechanisms and impacts of day

respiration in plant leaves: An introduction to a virtual issue. New Phytologist. 2021;**230**(1):5-10

[31] Howe Cheng T, Bc K, Uttraphan C, Mei Yee H. A review on energy harvesting potential from living plants: Future energy resource. International journal of Renewable Energy Research. 2018;**8**(4):2398-2414

[32] Ksenzhek OS, Volkov AG. Plant Energetics. Academic Press; 1998. pp. 133-154

[33] Xu Z, Jiang Y, Zhou G. Response and adaptation of photosynthesis, respiration, and antioxidant systems to elevated CO_2 with environmental stress in plants. Frontiers in Plant Science. 2015;**6**:1-17

[34] Wieloch T. A cytosolic oxidation–reduction cycle in plant leaves. Journal of Experimental Botany. 2021;**72**(12):4186-4189

[35] Ping MA, Tuan-hui BAI, Xiao-qian WANG, Feng-Wang MA. Effects of light intensity on photosynthesis and photoprotective mechanisms in apple under progressive drought. Journal of Integrative Agriculture. 2015;**14**(9):1755-1766

[36] Marouane B, Toshiaki M, Michael H, Eckart P, Herritt Matthew T, Iker A, et al. Photosynthesis in a changing global climate: Scaling up and scaling down in crops. Frontiers in Plant Science. 2020;**11**:1-29. DOI: 10.3389/fpls.2020.00882

[37] Males J. Secrets of succulence. Journal of Experimental Botany. 2017;**68**(9):2121-2134

[38] Griffiths H, Males J. Succulent plants. Current Biology. 2017;**27**(17):R890-R896

[39] Manvitha K, Bhushan B. *Aloe vera*: A wonder plant its history, cultivation and medicinal uses. Journal of Pharmacognosy and Phytochemistry. 2014;**2**(5):85-88

[40] Malik I. *Aloe vera*: A review of its clinical effectiveness. International Research Journal of Pharmacy. 2020;**4**:75-79

[41] Luttge U. Ecophysiology of crassulacean acid metabolism (CAM). Annals of Botany. 2004;**93**(6):629-652

[42] Bhardwaj T, Singh A, Agarwal D, Singh P. Generation of electricity from *Aloe vera* plant: A step towards creating an era. Foreword by Secretary. 2015;**47**:46-48

[43] Chong PL, Singh AK, Kok SL. Characterization of *Aloe barbadensis* Miller leaves as a potential electrical energy source with optimum experimental setup conditions. PLoS One. 2019;**14**(6):e0218758

[44] Chong PL, Singh AK, Kok SL. Potential application of *Aloe vera*-derived plant-based cell in powering wireless device for remote sensor activation. PLoS One. 2019;**14**(12):e0227253

[45] Koller AL, Rost TL. Leaf anatomy in *Sansevieria* (Agavaceae). American Journal of Botany. 1988;**75**(5):615-633

[46] Thiede J. *Beschorneria*. AGAVACEAE. 2019:1-9. DOI: 10.1007/978-3-662-56324-3_105-1

Chapter 4

Self-Powering Wireless Sensor Networks in the Oil and Gas Industry

Musaab Zarog

Abstract

The total revenue from the oil and gas industry in 2019 was 3 trillion dollars with nearly 350,000 businesses working in this field. For more efficiency, all machinery and equipment, including thousands of kilometers of transporting pipelines, need to be monitored continuously and in real time. Hundreds or even thousands of sensing and control nodes are needed for the oil and gas industry. WSNs approach has allowed the company to reduce the number of antenna towers and masts at remote sites, which accounts for 40–60% of the infrastructure cost of building a wireless digital oilfield network. A conventional solution to power these nodes is the use of electrochemical batteries. However, problems can occur using batteries due to their finite lifespan. The need for constant replacement in remote locations can become a very expensive or even impossible task. Over the last years, ambient energy harvesters have received great attention, including vibration-to-electric energy conversion. The aim of this chapter is to present the usefulness of implementing IoT and self-powered WSNs in the oil and gas sector, as well as challenges and issues related to adopting such a system.

Keywords: energy scavenging, wireless networks, sensors, MEMS, mechanical vibration, microsystems, ambient energy

1. Introduction

In the oil and gas industry, bulky wired cabling is not a good choice to monitor processes and communicate the information within the whole system. The energy industry is currently looking toward embracing IoT technology in almost all its operations, from monitoring well production to predicting when its gear will need maintenance. A recent report, produced by McKinsey Global Institute, estimates that $11.1 trillion a year in economic value by 2025 can be generated by moving from the physical world to the digital one [1]. The M2M (machine to machine) direct communication, between sensors and actuators through computing systems, can be achieved through the Internet of Things (IoT) and wireless sensor networks (WSNs). Today, many companies are developing wireless networks for various industrial applications, such as gas and oil industry. EE publishers produced a recent article in 2019 titled "Wireless monitoring to modernise the oil and gas industry" where it stressed industry 4.0 trends in the oil and gas industry through IoT and

WSNs and how these wireless technology can significantly affect the industry [2]. Wireless sensing technology is ideal for the oil and gas industry for many reasons, such as condition monitoring, production optimization, improving safety, and reducing the cost of wired devices [3]. In wireless sensing scenarios, hundreds or even thousands of sensors are deployed in a remote area, that is, production monitoring of an oil field, integrity monitoring of a long oil/gas pipeline infrastructure, or condition monitoring of a huge plant. There are many challenges faced while using conventional batteries to provide operating power to wireless sensing/control nodes [3]:

1. Limited lifetime of the batteries
2. Need to continuously replace batteries at thousands of points.
3. Batteries replacement could be a very time-consuming task and even uneconomical and unmanageable. In some applications, replacing batteries is not practical.
4. Huge maintenance effort would be required to replace or recharge the batteries of these sensors.
5. Sensors' battery replacement for pipelines buried in soil or water, or in a hazardous environment that could require the shutdown of a plant and operation.
6. In addition, in remote and difficult-to-access locations such as subsea oil fields, battery replacement or recharging could be very expensive.
7. The performance and reliability of conventional batteries (primary/rechargeable) drastically degrade in harsh environments, which are very common in the oil and gas industry.
8. Relatively big size of battery compared to other devices in the node (e.g. sensors and actuators)

Self-powering devices can resolve all the previously mentioned issues completely.

The pipeline infrastructure of thousands of kilometers also possesses a very small magnitude of vibrations at the pipeline surface. The pipeline carrying liquid (oil and water), gas, or a multiphase flow can exhibit vibrations. The nature of flow-induced vibration in a pipe conveying fluid is a broadband frequency vibration [4]. The turbulence-induced vibration generates random pressure fluctuations around the inner surface of the pipe forcing it to vibrate. In the case of plants and refineries, line-powered machinery is an excellent vibration source to harvest from. They have a repeatable frequency component of 50 or 60 Hz (typical line power frequency). Mechanical energy harvesting techniques can be used to convert mechanical vibrations to electrical energy. One example is a piezoelectric- or electromagnetic-based energy harvester, tuned at the structural vibration frequency of pipeline or process equipment. The harvester should have sufficient bandwidth and be able to operate at a range of frequencies. The power generated by the harvester can be utilized for many sensing applications including equipment condition monitoring, pipeline integrity monitoring, and production monitoring. **Figure 1** shows pressure, flow, temperature,

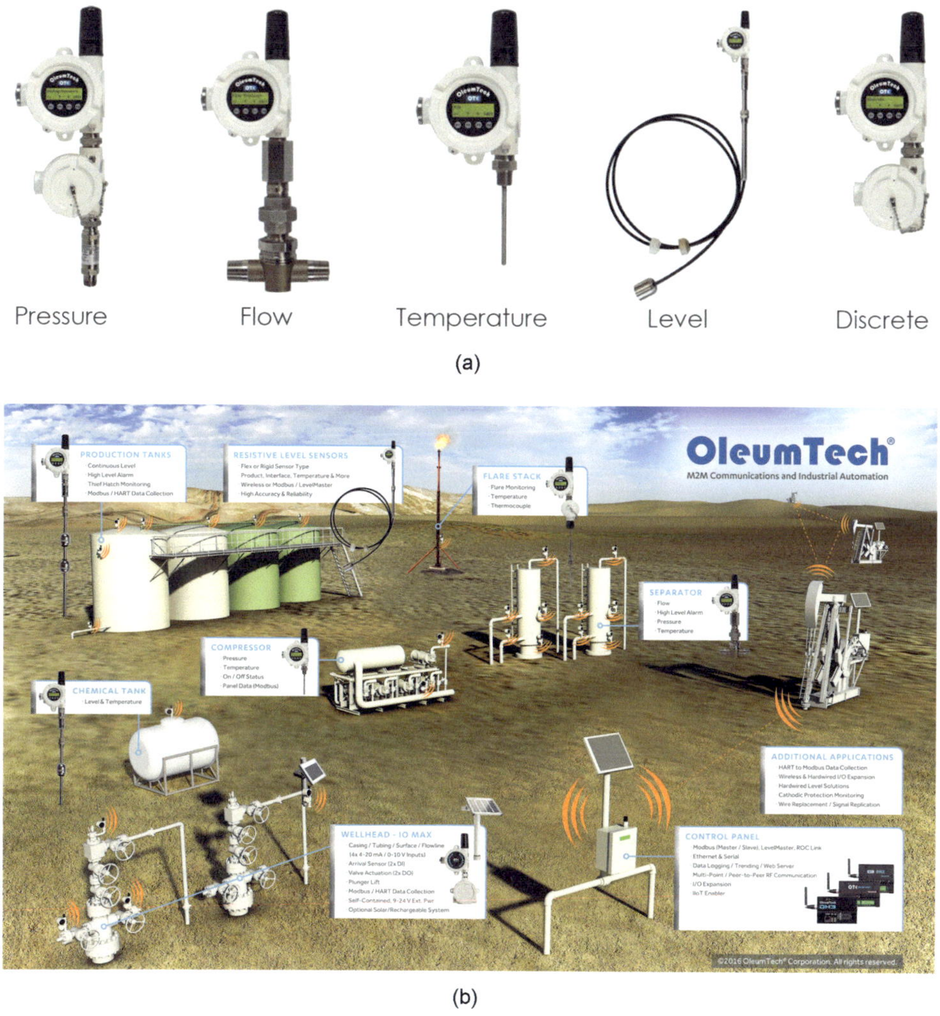

Figure 1.
(a) Example of sensor nodes used in oilfields (b) and their location in oil and gas field [4].

and level sensors that gather important industrial data and their location within the oil and gas industry [4].

2. Self-powered systems and sensors

Nowadays, there is a global need for sustainable energy generation, storage, and distribution of infrastructure to face challenges in developing energy technologies that will provide the amount of energy needed on a sufficient scale and timeframe with minimal impact on the environment and have limited economic and societal disruption during implementation [5]. There have been continuous efforts and tremendous interest to harvest ambient mechanical energy. Therefore, with the increasing demand for electrical energy, energy conversion from renewable energy resources has received huge attention in the last few decades. For low-power electronic devices (e.g. wireless network sensors), continuous replacement of these portable electronic batteries can be problematic and time consuming. Current average global primary power consumption sits at approximately 14 terawatts (TW, 14 × 1012 Jouless−1), with more than 80% of this energy coming from the carbon-dioxide emitting fossil fuel trio of oil, coal, and natural gas, and less than 1% coming from carbon-free renewable power, such as geothermal, wind, solar power, and biofuels [5].

The conversion of waste heat to electrical power arguably represents the greatest opportunity for energy conservation [5]. One promising approach to improve energy efficiency is to employ solid-state thermoelectric (TE) devices to recover part of this waste heat and convert it directly into electricity by utilizing the Seebeck effect [5]. Triboelectric nanogenerators (TENGs) facilitate an excellent opportunity to power wireless sensors and systems that requires few milliwatts of power which can be built into these sensors and devices. The working principle of TENGs is based on contact electrification and electrostatic induction to harvest waste mechanical energy. Therefore, there are plenty of ambient sources, such as wind energy, water wave energy, and vehicle and human body motion, that can be used to extract the energy using TENGs principle. There are successful attempts to extract biomechanical energy using TENGs and implement the harvested energy to power devices, such as touch pads, health monitoring systems, security application systems, exoskeletons, gloves, and many other portable and wearable electronics. The suitability of TENGs was further emphasized by their high adaptability, simple design, ease of fabrication, and cost-effectiveness. Nevertheless, improving power generation remains the main focus of the current research. For example, Dudem reported TENG energy-harvesting unit, that generates a maximum electrical output current of 3.5 μA and power of 0.61 mW, under an applied pushing frequency and force of 4 Hz and 25 N [6].

Piezoelectric nanogenerators (PNGs) are one of the promising technologies for energy harvesting. PNGs utilize piezoelectric materials that are capable of converting mechanical energies into electricity. When piezoelectric materials are stressed, electric charges accumulate at the surface, where they produce electricity that can be used to power portable electronic devices. The piezoelectric effect is also reversible, so that when electric load is applied, the piezoelectric material deforms. Piezoelectric nanogenerators can generate power density that is comparable with the power density of solar cells. For a large piezoelectric effect, the material needs to be initially polarized by applying high voltages across the piezoelectric material. The piezoelectric effect can be generally observed in inorganic materials, such as quartz, lead zirconium titanate (PZT; Pb(Zr,Ti)O3), zinc oxide (ZnO), and barium titanate (BTO, BaTiO3).

Under a very small dynamic force (less than 0.06 N), the output power of 1.5 mW was obtained with an 8.5 mm drum harvester across a load resistance of 17.8 kΩ at a frequency of 173 Hz [7]. TENGs were also combined with other energy-harvesting technologies, such as solar cells and electromagnetic generators, yielding hybrid nanogenerators that are able to operate in broader operating frequency ranges. Besides, the piezoelectric materials were also used in combination with the triboelectric polymers as active layers to form hybrid devices with enhanced output performances [8].

3. IoT for oil and gas industry

The Internet of Things (IoT) is a set of physical objects or devices that are capable of exchanging data over a communication network where sensors play an essential role in reading physical information. IoT facilitates continuous interaction between objects and devices (things). There are many interesting applications in many different areas, such as scientific, commercial, civil, and military domains, for continuous event detection and monitoring. Examples of this application are building management (smart homes/offices), smart manufacturing healthcare monitoring, smart cities, environment monitoring, tracking, security, and surveillance. **Figure 2** shows the function cycle of the Internet of Things [9]. A typical WSN-based IoT setup is formed by a number of distributed and autonomous sensing devices or nodes. These sensing networks need power to operate and that necessitate continuous replacement of batteries. Self-powering systems will be an ideal solution for wireless sensor networks [10].

Energy can be extracted from external sources through the process of energy harvesting or power harvesting, to be used for any viable purposes. External sources include solar power, thermal energy, wind energy, and kinetic energy, also known as ambient energy. The universal energy crisis and the global trend to use renewable energy have boosted research in the area of energy harvesting. One of the newly emerging areas is harvesting power for wireless autonomous devices. Nowadays, with the increase in power consumption, we continuously see electronic devices getting smaller in this era Internet of Things (IoT). According to a report published by Statista Research Department in November 2019, the total installed base of Internet of Things (IoT) connected devices is projected to be 75.44 billion worldwide by 2025 [11]. All these IoT devices face a real battery life problem where batteries need to be continuously charged or replaced. Energy harvesters provide a solution to the IoT battery problem. IoT is just an example of factors that have driven the energy harvesting market to grow at a large scale. According to a press release published in August 2019 by Market Watch (which is a leader in providing the latest financial news and market data), the major driving factor for the need for energy harvesting is the growing adoption of energy harvesting systems in wireless sensor networks (**Figure 3**) [12].

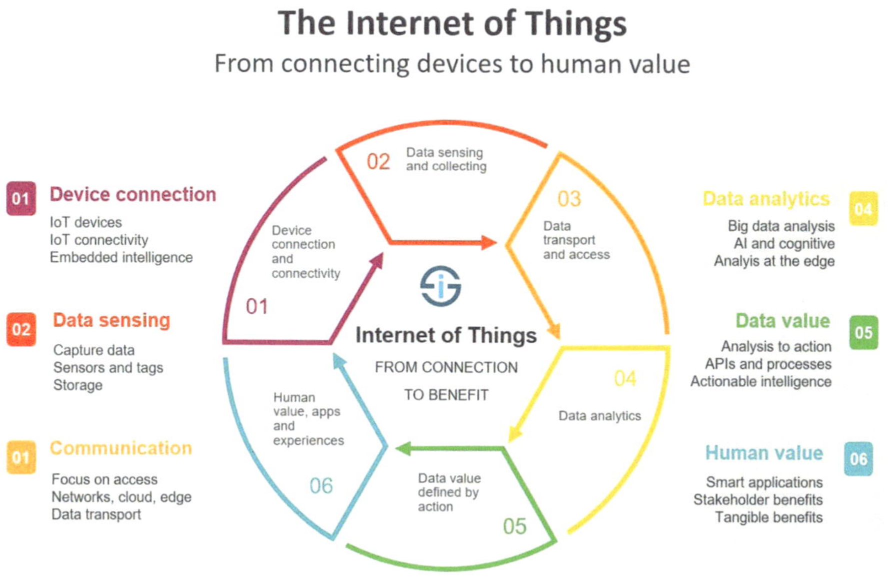

Figure 2.
Internet of things: Function cycle [9].

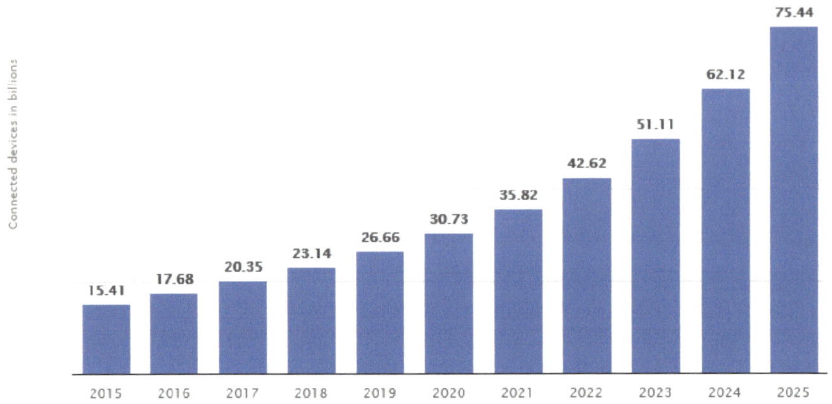

Figure 3.
Internet of things (IoT) connected devices from 2015 to 2025 (in billions) [11].

The energy harvesting chip market is expected to approach $3.4 bn by 2022 [13]. The gas and oil industry takes place in remote areas that need to be monitored continuously, besides, it requires remote monitoring of pipelines, gas and oil leaks, carrion, and equipment conditions all through upstream, midstream, and downstream. Wired technology is expensive, difficult to maintain, and less capable of working in a harsh environment. Wireless technology proved to be the ultimate solution to monitor and control oil and gas industries. Failures in pipelines have led to an annual loss of up to 10 billion USD in the United States, environmental disasters, and fatal incidents [14]. A Market Dynamics Report produced by ON World shows that oil and gas wireless sensor network is one of the key technologies that are transforming the oil and gas industry. According to their survey of 110 leading firms in the oil and gas industry, the majority of these firms are moving toward WSN solutions [4]. Some of these firms have already deployed more than 1000 nodes in their oil and gas industry. Millions of WSN devices will be deployed worldwide for well automation, pipeline operation, and exploration of new and existing oilfields. Oil and gas companies are adopting wireless sensor instruments that provide up to 80% infrastructure savings compared with wired options [11]. According to the market research report, the wireless sensor network (WSN) market was valued at USD 29.06 billion in 2016 and is expected to reach USD 93.86 billion by 2023 [15]. In 2023, global WSN revenues for oil and gas exploration, production, and pipeline operation will reach $2.2 billion. Oil and gas companies are adopting wireless sensor instruments that provide up to 80% infrastructure savings compared with wired options [16]. ON World's Q2 2018 survey found that two in five of the oil and gas respondents have installed over 1000 WSN nodes across all locations. Twenty percent (20%) have deployed networks with at least 3000 nodes compared with 6% in our previous survey in Q4 2016. ON World's survey found that the fastest-growing oil and gas WSN applications are asset tracking and locating, process control, environmental/safety, and asset/machine health monitoring [17].

There is a continuous growth of research for self-powered WSNs that can be used to monitor machines, processes via sensing different parameters (such as temperature, vibration, strain, rotating speed, displacement, pressure, voltage, current, and acoustics) [18]. Besides monitoring machines in the oil and gas industry, WSNs can also be used to monitor pipelines condition [19–22].

4. MEMS energy harvesters for oil and gas: options and challenges

There is continuous progress in reducing the power needed for wireless electronics for wireless sensors, which is now in the range of milliwatts. The milliwatts that need to power wireless sensing and control still pose an issue since they have to be supplied by batteries that need continuous replacement. Hundreds of thousands of network sensing batteries are to be placed along the pipeline and all over the huge oil and gas plants. Replacement or charging is costly and time-consuming, and it gets more complicated for sensing networks that are buried in soil, underwater, or exist in a hazardous environment. On the other hand, the disposal of these huge number of batteries is another environmental issue. An ideal solution to the aforementioned issues, (related to cost, safety, and environment) is to use self-powered devices that can extract free and renewable energy from the surrounding environment. With the advancement in microelectromenchal systems technology (MEMS), the energy harvester devices can be miniaturized to a very small scale and produce power in the range of microwatts. The energy can be harvested from ambient vibration that exists in oil and gas equipment, machines, and pipelines. The most common principles of harvesting vibration energy are electromagnetic, electrostatic [23], and piezoelectric [24–26]; while the latter has many advantages over other principles, since it does not need a voltage source like in the case of electrostatic harvesters and at the same time can be downscaled in size easily as opposed to the electromagnetic harvesters. Piezoelectric micro harvesters have proved to have the capability of producing large voltage output compared to others [27]. Although piezoelectric MEMS have proved to be good candidates for micro energy harvesting, they still face some challenges (e.g. small current output and depolarization), and research is being carried out to resolve these issues.

One of the first attempts to harvest energy from athletic shoes was carried out in 1998 [28]. The shoe contains piezoelectric material to convert dynamic force to electricity as an example of wearable power-generating systems. Although the whole system was bulky, this research has imitated more research to be carried out to improve circuit design as well as to miniaturize the system further. With the currently advancing MEMS technologies, research is now focusing a lot on MEMS-based energy harvesters. One issue, which was discussed earlier, is the electromechanical coupling factor for piezoelectric materials and its contribution to the produced power.

Typical modes of vibration discussed in the literature are mode31 and mode33, which mainly depend on the direction in which we are applying the force to the piezoelectric structure (vibration force being applied perpendicular to the polling direction in the case of mode31 or applied horizontally in the mode33). It was found that the mode33 produces larger electromechanical coupling which indicates higher generated power. Since power generated from piezoelectric materials depend on both the coupling factor and the amplitude of the applied mechanical stress, it was found that the mode31 produces larger mechanical stresses which imply larger electric power. Since the advantage of high mechanical stress overcomes the advantage of high coupling effect [29], mode31 is more preferred and hence adopted in the literature. The simplest structure that can produce high mechanical stresses is the beam-like structure (cantilevers and bridges) with cantilevers being simpler and they suffer less

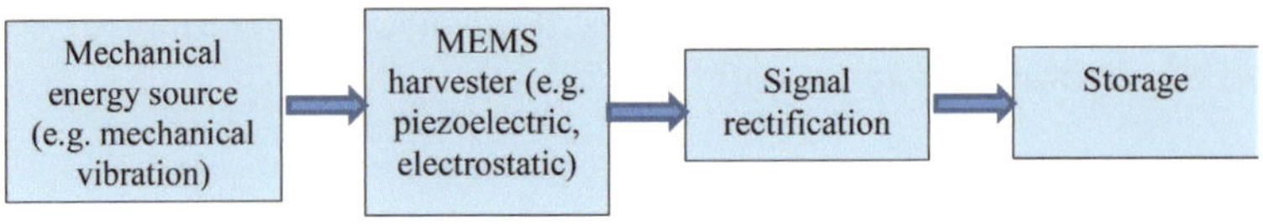

Figure 4.
Schematic diagram of typical MEMS harvesting system.

from other issues, such as residual stresses, developed in the structure. Therefore, the cantilever-like structure is more utilized for piezoelectric energy harvesting. Since there are many types of piezoelectric materials which can be used to harvest energy, a lot of research was made to investigate the advantages and disadvantages of piezoelectric materials that can be used to harvest mechanical energy. The most common piezoelectric thin films used for energy harvesting are PZT, PMN-PT, KNN, BiFeO3, BT, ZnO, AlN, and PVDF. Research indicates that PZT overcomes other candidates in terms of the piezoelectric coefficient [19], but PZT suffers from the issue of brittleness. It is well known that the highest extracted power from these harvesters can be achieved at the resonance frequency of the structure. Different structure layouts will result in different resonance frequencies. Another issue related to the level of power harvested is tuning the energy harvester device to the ambient vibration frequency. This issue was dealt with at two levels; lowering the resonance frequency of the device to match the ambient vibration and the other level reshaping the architecture and dimensions of the device to match the most common ambient vibration frequency. Shape, architecture, and sizing of the harvester device have been investigated intensively in the literature [30–35]. Among the top issues considered in recent research is lowering the resonant frequency of these MEMS micro harvesters to match the ambient vibration and conditions (e.g. acceleration, level of vibration, and frequency range) in the oil and gas industry [7, 36]. Another challenge is widening the bandwidth of these micro harvesters [37–42]. Different MEMS designs and solutions were proposed to overcome these challenges while maximizing the output power of these micro harvesters [43–54]. There is still more to do to overcome challenges related to the optimization of power harvested through MEMS devices, and how to increase the broadband capability of the MEMS harvesters while optimizing the power generated from the device. The piezoelectric harvesters and electrostatic harvesters are more feasible than electromagnetic harvesters to their down-scalability and possibility of fabricating MEMS energy harvesters, with the preference for piezoelectric harvesters due to their higher power intensity over electrostatic harvesters. **Figure 4** represents a typical MEMS harvesting system.

5. Recent trends in applying IoT & energy harvesting in oil and gas industry

The operations of the oil and gas industries present a huge amount of data. In the midstream sector alone, every 150,000 miles of pipeline produces up to 10 terabytes of data [55, 56]. More recent approaches to monitor pipeline networks are based on wireless sensor networks (WSN) and Internet of Things (IoT) for the three sectors of the oil and gas industry [14]. Today, there are several companies, such as Ambyint (previously PumpWell) and WellAware, that provide IoT solutions for wellhead and pump jack monitoring and other gas and oil monitoring needs, that are implemented.

According to WellAware, its production management solutions help companies reduce lease-operating expenses, minimize unplanned downtime, and ensure safety and regulatory compliance [57]. The diagram below describes some of the common threats, to good performance, which can be overcome by implementing IoT (**Figure 5**) [57]. BehrTech, a provider of wireless IoT connectivity for industrial and commercial networks, identified five areas of applications of IoT for the oil and gas industry, 1) asset maintenance, 2) hazard management and worker safety, 3) facility management, 4) regulatory compliance, and 5) security and access control [58]. Brian Ray, the founder and CTO of Link Labs, identified four areas, where connected wireless technology is helping the oil and gas industries to 1) optimize for efficient pumping activities, 2) maintain the pipes and wells, 3) monitor equipment failures and gas leaks, and 4) monitor pipe thickness, temperatures, and erosion in a refinery [57]. ON World Research, leading industry experts reported that oil and gas companies are adopting wireless sensor instruments that provide up to 80% infrastructure savings compared with wired options. They also report that, as oil prices continue to rise and

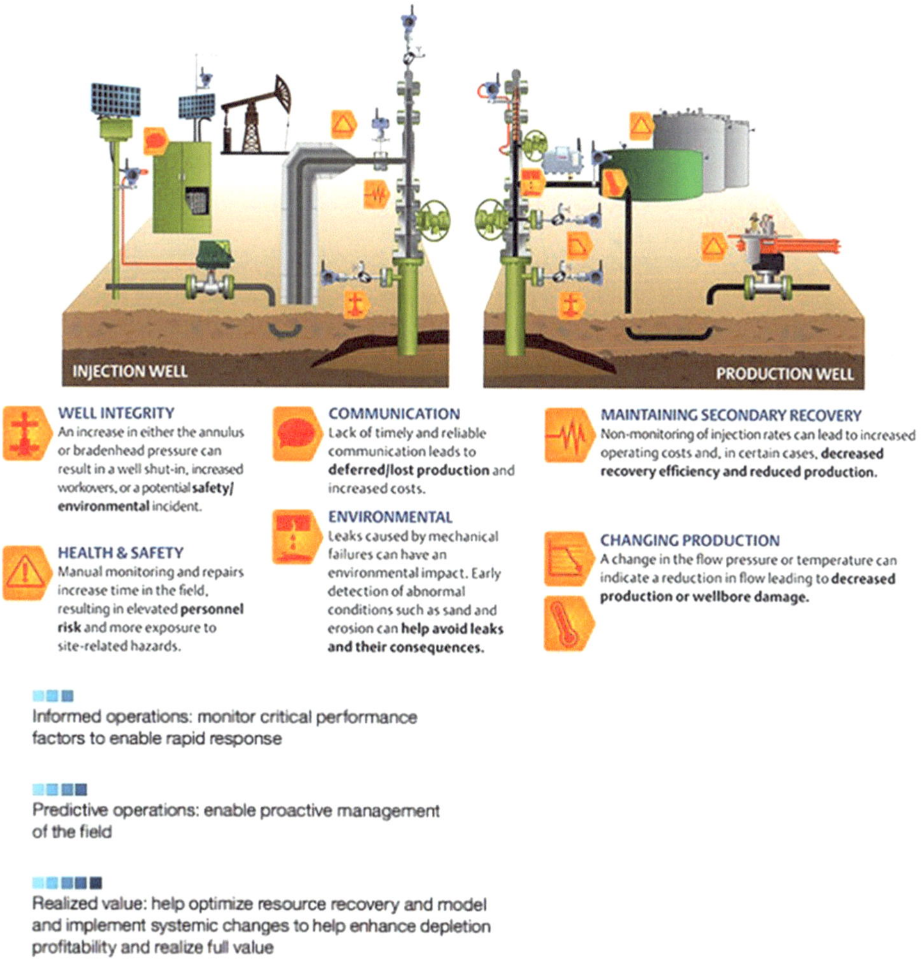

Figure 5.
Common threats to onshore well performance which can be overcome by implementing IoT [57].

exploration activity increases, WSN adoption is steadily growing for core applications, such as wellhead automation and pipeline compressor/pump station monitoring as well as growing innovations for asset management, worker safety, and environmental monitoring. They are expecting that in 2023, global WSN revenues for oil and gas exploration, production, and pipeline operation will reach $2.2 billion up from $480 million in 2017 [16]. BP (British Petroleum), which is one of the leading names in this space, has teamed up with GE (General Electric) to connect its oil rigs to the internet. The collected information can be analyzed in real time and uploaded to the cloud, where trained engineers can make further analyses [59]. Hill Corp, a famous energy company in the USA, uses sensors throughout the pumping system which are then fed into the Microsoft Azure Cloud, from where data is pushed to engineers working on digital dashboards who can monitor the pump's health and performance. The estimates are that a failure of its pumps could cost the company up to $300,000 a day in lost production [14]. Petroleum Development Oman (PDO) oil field covers an area of 72,520 sqkm (28,000 square miles) of desert, rugged, resilient, and reliable equipment is critical. In Oman, over 2000 of the 5000 oil wells in their Oman location are now connected wirelessly using redline equipment to a centralized site, where they are monitored and managed remotely in real time, eliminating the need to have workers drive from well-to-well to collect critical operating information and make changes to optimize the system. PDO is now expanding wireless coverage to all wells. SCADA systems, RTUS, video surveillance cameras, and Wi-fi hotspots are all connected over the wide-area Redline network. This WSNs approach has allowed the company to reduce the number of antenna towers and masts at remote sites, which account for 40–60% of the infrastructure cost of building a wireless digital oilfield network [60].

Having introduced the importance of IoT applications in the oil and gas industry, which can only be achieved by the installation of wireless networks, the challenge now is how to power these WNs in the oil and gas remote areas. Energy harvesters provide a solution to the IoT battery problem. Changing batteries for thousands of remotely deployed wireless sensor nodes could become an expensive logistical headache. ON World Research, leading industry experts reported that, energy harvesting was identified as the most needed innovation for wireless sensor networks. However, it is difficult to power wireless sensor nodes directly from solar panels since supply voltage depends on the time-varying load impedance [16]. A promising alternative to solar panels is mechanical vibration energy harvesting. Ahmad Talha and his colleagues at ARAMCO Saudi, in their conference paper titled "Energy harvesting, powered wireless monitoring and control in oil and gas," have demonstrated the great potentials and promising opportunities of harvesting energy to power wireless networks in the oil and gas industry [61]. They point out that energy can be harvested from this excessive flow energy and can be used for a variety of applications since the fluid flow is present almost everywhere inside pipelines carrying fluid (oil, gas, or multiphase) from well sites to refineries and then to seaports. There were successful attempts to generate the required energy to power WNs devices for IoT. Honeywell, diversified technology and manufacturing company, has developed wireless industrial transmitters (e.g. XYR6000) powered with thermoelectric energy harvesters, and this solution has addressed a key customer objection about batteries and expanded the environments where wireless is the best choice [62]. Perpetuum, a global leading company in wireless sensing technology, produced vibration energy harvesters to powered wireless sensor nodes used to monitor valuable equipment and processes within the oil and gas industry. They claim that the self-powered industrial wireless sensor nodes (WSN) can work over 10 years without changing batteries [14]. ReVibe Energy, a company founded to power the industrial Internet of Things, has developed three vibration energy harvesting products, to provide wireless power for sensor systems in the industry such as process manufacturing, oil and gas, railways, mining, and aerospace [63, 64]. The products vary in terms of the range of frequency of operation, acceleration range, and expected range of produced power.

6. Conclusions and recommendations

Implementing wireless technology and IoT can result in a huge reduction in the cost of monitoring and controlling the gas and oil industry (e.g. failures in the oil and gas industry and hence reduce the shut-down time). It also provides a real time managing system as opposed to traditional walk-through systems. At this stage, it is clear that there are many issues needed to be tackled to enable fully autonomous and maintenance-free wireless sensors for various applications in the oil and gas industry, as well as to provide the oil and gas industry with a continuous source of energy. Energy harvesting from ambient vibration existing in the oil and gas industry can result in further reduction in cost, inconvenience, and efforts, to replace batteries that power the wireless sensor networks (WSNs) and eliminate the use of chemical batteries and their negative environmental impact after disposal. To optimize the vibrational energy harvesting in the gas and oil industry, one must consider the level and frequency of vibration that exist in the fields. More investigation of the ambient mechanical vibration of equipment and pipelines in the oil and gas industry (e.g. level and frequency of vibration and the broadband range) is required. Implementing MEMS solution for power harvesting has another advantage massive production and subsequent cost reduction. Finally, despite the current efforts, more research is needed to design and implement a self-powered wireless system to monitor and control the oil and gas industry.

Author details

Musaab Zarog
Department of Mechanical and Industrial Engineering, College of Engineering, Sultan Qaboos University, Muscat, Oman

*Address all correspondence to: musaabh@squ.edu.om

References

[1] McKinsey Global Institute. https://www.mckinsey.com/business-functions/mckinsey-digital/our-insights/the-internet-of-things-the-value-of-digitizing-the-physical-world

[2] Adejo AO, Onumanyi AJ, Anyanya JM, Oyewobi SO. Oil and gas process monitoring through wireless sensor networks: A survey. Ozean Journal of Applied Science. 2013;**6**(2):39-43

[3] Ahmad TJ, Noui Mehidi MN. Energy harvesting powered wireless monitoring and control in oil and gas. In: SPE Middle East Intelligent Energy Conference and Exhibition held in Dubai. UAE; 2013

[4] OleumTech. https://www.prnewswire.com/news-releases/oleumtech-wireless-sensor-networks-are-enabling-oil--gas-operators-to-increase-efficiency-and-reliability-while-cutting-costs-300319843.html

[5] Graham SA, Dudem B, Mule AR, Patnam H, Yu JS. Engineering squandered cotton into eco-benign microarchitectured triboelectric films for sustainable and highly efficient mechanical energy harvesting. Nano Energy. 2019;**61**:505-516

[6] Bhaskar Dudem RD, Dharmasena IG, Riaz R, Venkateswaran Vivekananthan KGU, Wijayantha PL, Petti L, et al. Wearable triboelectric nanogenerator from waste materials for autonomous information transmission via Morse code. ACS Applied Materials & Interfaces. 2022;**14**(4):5328-5337

[7] Zarog M. Piezoelectric ceramic for energy harvesting from ambient vibration. Energetika. 2021;**67**:1-2

[8] Adonijah Graham S, Dudem B, Patnam H, Mule AR, Yu JS. Integrated design of highly porous cellulose-loaded polymer-based triboelectric films toward flexible, humidity-resistant, and sustainable mechanical energy harvesters. ACS Energy Letters. 2020;**5**(7):2140-2148

[9] Christodoulou S. ICC Property Management: How Technology Transforms Property Management Industry. 2020. https://www.stevenchristodoulou.com/icc-property-management-how-technology-transforms-property-management-industry/

[10] Haq IU, Javaid Q, Ullah Z, et al. E2-MACH: Energy efficient multi-attribute based clustering scheme for energy harvesting wireless sensor networks. International Journal of Distributed Sensor Networks. 2020;**16**(10):1-7

[11] https://www.statista.com/statistics/471264/iot-number-of-connected-devices-worldwide, A report published in November 2019

[12] Energy Harvesting System Market 2019|Top Key Players Analysis, Trends, Global Size Forecast to 2024, A press release, August 2019, https://www.marketwatch.com/press-release/energy-harvesting-system-market-2019top-key-players-analysis-trends-global-size-forecast-to-2024-2019-08-12

[13] The Economist, Technology Quarterly Sep 2019 edition, https://www.economist.com/technology-quarterly/2019/09/12/how-to-build-a-disposable-microchip

[14] PERPETUUM, https://perpetuum.com/industrial-applications/

[15] MarketsandMarkets, Wireless Sensor Network Market worth 93.86 Billion

USD by 2023, press release, https://www.marketsandmarkets.com/PressReleases/wireless-sensor-network.asp

[16] ON World's, https://onworld.com/news/2-Billion-Digital-Oilfield-Wireless-Sensor-Network-Market-in-2018.html

[17] San Diego (PRWEB). Bottom Line Focus Will Drive Oil and Gas Wireless Sensor Network Revenues to $2.2 Billion in 2023. 2018. https://www.prweb.com/releases/bottom_line_focus_will_drive_oil_and_gas_wireless_sensor_network_revenues_to_2_2_billion_in_2023/prweb15806653.htm

[18] ON World's report, Oil & Gas Wireless Sensor Networks, A Market Dynamics Report, 6th edition, https://www.onworld.com/oilandgas/

[19] Tang X, Wang X, Cattley R, Gu F, Ball AD. Energy harvesting technologies for achieving self-powered wireless sensor networks in machine condition monitoring: A review. Sensors. 2018;**18**:4113

[20] Zhu J et al. Gas pipeline leakage detection based on PZT sensors. Smart Materials and Structures. 2017;**26**:025022

[21] Siswantoro N, Doğan A, Priyanta D, Zaman MB, Semin S. Possibility of piezoelectric sensor to monitor onshore pipeline in real time monitoring. International Journal of Marine Engineering and Innovative Research. 2019;**3**:199-211

[22] Hou Q, Zhu W. An EKF-based method and experimental study for small leakage detection and location in natural gas pipelines. Applied Sciences. 2019;**9**(15):3193

[23] Pop-Vadean A, Pop PP, Barz C, Chiver O. Applications of energy harvesting for ultralow power technology. IOP Conference Series: Materials Science and Engineering. 2015;**85**:012024

[24] Hassan M. Determination of residual stresses in a single crystalline 3C-SiC micro-fabricated structure using FE model and measured resonance frequencies. Journal of Microsystem Technologies. 2012;**18**(3)

[25] Tao K, Lye SW, Wang N, Hu X, Miao JM. A sandwich-structured MEMS electret power generator for multi-directional vibration energy harvesting. In: 2015 Transducers - 2015 18th International Conference on Solid-State Sensors, Actuators and Microsystems (TRANSDUCERS). Anchorage, AK; 2015. pp. 51-54

[26] Muscalu G, Anghelescu A, Firtat B. Design optimization of MEMS piezoelectric energy cantilever device for environment vibrations harvesting. In: 2015 International Semiconductor Conference (CAS). Sinaia; 2015. pp. 267-270

[27] Rana VS, Chauhan A. Energy harvesting from locomotive and coaches by piezoelectric ceramic using MEMS technology. In: 2015 International Conference on Innovations in Information, Embedded and Communication Systems (ICIIECS). Coimbatore; 2015. pp. 1-6

[28] Kymissis J, Kendall C, Paradiso J, Gershenfeld N. Parasitic power harvesting in shoes. In: Proc. of Second International Symposium on Wearable Computers. Pittsburgh; 1998. pp. 132-139

[29] Matiko JW, Beeby S. Applications of Energy Harvesting Technologies in Buildings. Artech House; 2017. p. 170

[30] Todaro MT, Guido F, Mastronardi V, Desmaele D, Epifani G, Algieri L, et al. Piezoelectric MEMS vibrational energy harvesters: Advances and outlook. Microelectronic Engineering. 2017;**183-184**:23-36

[31] Saadon S, Sidek O. Shape optimization of cantilever-based MEMS piezoelectric energy harvester for low frequency applications. In: UK Sim 15th International Conference on Computer Modelling and Simulation. Cambridge, UK; 2013

[32] Rosmi AS, Isa M, Ismail B, Rohani MNKH, Wahab Y. An optimization of electrical output power for piezoelectric energy harvester using different micro-Cantilever beam geometries. In: AIP Conference Proceedings. 2017

[33] Mohamed R, Sarker MR, Mohamed A. An optimization of rectangular shape piezoelectric energy harvesting cantilever beam for micro devices. International Journal of Applied Electromagnetics & Mechanics. 2016;**50**(4)

[34] Benasciutti D, Moro L, Zelenika S, a., & Brusa, E. Vibration energy scavenging via piezoelectric bimorphs of optimized shapes. Microsystem Technologies. 2010;**5**:657

[35] Luo Y, Gan R, Wan S, Xu R, Zhou H. Design and analysis of a MEMS-based bifurcate-shape piezoelectric energy harvester. AIP Advances. 2019;**6**(4)

[36] Ramírez JM, Gatti CD, Machado SP, Febbo M. A piezoelectric energy harvester for rotating environment using a linked E-shape multi-beam. Extreme Mechanics Letters. 2019;**27**:8-19

[37] Chopra K, Pandey S. A new design of Aluminium cantilever with embedded piezoelectric ceramic film in RF MEMS devices for energy harvesting. In: TENCON 2015 - 2015 IEEE Region 10 Conference. Macao; 2015. pp. 1-6

[38] Rezaei M, Khadem SE, Firoozy P. Broadband and tunable PZT energy harvesting utilizing local nonlinearity and tip mass effects. International Journal of Engineering Science. 2017;**118**:1-15

[39] Usharani R, Uma G, Umapathy M, Choi S-B. Design of high output broadband piezoelectric energy harvester. Journal of Mechanical Science and Technology. n.d.;**31**(7):3131-3142

[40] Usharani R, Uma G, Umapathy M. Design of High Output Broadband Piezoelectric Energy Harvester with double tapered cavity beam. International Journal of Precision Engineering and Manufacturing-green Technology. n.d.;**3**(4):343

[41] Elahi H, Eugeni M, Gaudenzi P. A review on mechanisms for piezoelectric-based energy harvesters. Energies. n.d.;**11**(7)

[42] HL-C, Deng TZ. Energy harvesting from low frequency applications using piezoelectric materials. Applied Physics Reviews. 2014;**1**(4)

[43] Tang G, Yang B, Hou C, Li G, Liu J, Chen X, Yang C. A Piezoelectric micro Generator Worked at Low Frequency and High Acceleration Based on PZT and Phosphor Bronze Bonding. 2016

[44] Berdy D, Srisungsitthisunti P, Jung B, Xu X, Rhoads J, Peroulis D. Low-frequency meandering piezoelectric vibration energy harvester. IEEE Transactions on Ultrasonics, Ferroelectrics and Frequency Control. 2012;**59**(5):846-858

[45] Jia Y, Seshia AA. Power optimization by mass tuning for MEMS piezoelectric cantilever vibration energy harvesting. Journal of Microelectromechanical Systems. Feb 2016;**25**(1): 108-117

[46] Tang G, Liu J, Yang B, Luo J, Liu H, Li Y, et al. Fabrication and analysis of high-performance

piezoelectric MEMS generators. Journal of Micromechanics and Microengineering. 2012;**22**(6)

[47] Lin S-C, Wu W-J. Fabrication of PZT MEMS energy harvester based on silicon and stainless-steel substrates utilizing an aerosol deposition method. Journal of Micromechanics and Microengineering. 2013;**23**(12):125028

[48] Lueke J, Rezaei M, Moussa W. Investigation of folded spring structures for vibration-based piezoelectric energy harvesting. Journal of Micromechanics and Microengineering. 2014;**24**(12)

[49] Wang X, Chen C, Wang N, San H, Yu Y, Halvorsen E, et al. A frequency and bandwidth tunable piezoelectric vibration energy harvester using multiple nonlinear techniques. Applied Energy. 2017;**190**:368-375

[50] Liu H, Tay CJ, Quan C, Kobayashi T, Lee C. Piezoelectric MEMS energy harvester for low-frequency vibrations with wideband operation range and steadily increased output power. Journal of Microelectromechanical Systems. 2011;**20**(5):1131-1142

[51] Shang Z, Cheng Y, Zhu Y. A micromachined low-frequency piezoelectric harvester for vibration and wind energy scavenging. Journal of Micromechanics and Microengineering. 2013;**23**(12)

[52] Jia Y, Du S, Seshia AA. Micromachined cantilevers-on-membrane topology for broadband vibration energy harvesting. Journal of Micromechanics and Microengineering. 2016;**26**(12)

[53] Truong BD, Phu Le C, Halvorsen E. Power optimization and effective stiffness for a vibration energy harvester with displacement constraints. Journal of Micromechanics and Microengineering. 2016;**26**(12)

[54] Nabavi S, Zhang L. Portable wind energy harvesters for low-power applications: A survey. Sensors. 2016;**16**(7)

[55] Deloitte Insights. Connected Barrels: Transforming Oil and Gas Strategies with the Internet of Things. The Internet of Things in the Oil and Gas Industry; 2015. Available from: https://www2.deloitte.com/us/en/insights/focus/internet-of-things/iot-in-oil-and-gas-industry/

[56] Ahmed S, Le Mouël F, Stouls N. Resilient IoT-based monitoring system for crude oil pipelines. In: 2020 7th International Conference on Internet of Things: Systems, Management and Security (IOTSMS). 2020

[57] Ray B. IoT In oil & gas: Analyzing Technology & use Cases. 2018, https://www.link-labs.com/blog/iot-oil-gas-use-cases]

[58] IoT Applications for Offshore Monitoring in Oil and Gas, https://behrtech.com/blog/5-iot-applications-for-offshore-monitoring-in-oil-and-gas/]

[59] Next Generation Sensor Network Solutions for Offshore Oil and Gas Operations, special report, Published by Global Business Media, Mojix Inc, published on Nov 13, 2015

[60] Oman Daily Observer, Canadian RM wins wireless contract covering 2,000 PDO wells, Report published in 10th April 2012

[61] Ahmad TJ, Noui MN, Mehidi. Energy harvesting powered wireless monitoring and control in oil and gas. In: Paper Presented at the SPE Middle East Intelligent Energy Conference and Exhibition. Manama, Bahrain; 2013

[62] Bragg M. Honeywell presentation: ENERGY HARVESTING in the OIL & GAS sector, 12th March 2014, chrome-extension:// efaidnbmnnnibpcajpcglclefindmkaj/ viewer.html?pdfurl=http%3A% 2F%2Feh-etwork.org%2Fevents%2Feh20 14%2Fspeakers%2F3.1.pdf&clen=915103 8&chunk=true

[63] ReVibe Energy, https://revibeenergy. com/

[64] Baxter J, Bian Z, Chen G, Danielson D, Dresselhaus M, Fedorov A, et al. Nanoscale design to enable the revolution in renewable energy. Energy & Environmental Science. 2009;**2**(6):559

Chapter 5

Solar Photovoltaic Energy System

Sanusi Yekinni, Ibrahim Asiata, Oyeshola Hakeem and Lawal Mubarak

Abstract

Solar Energy in term of Photovoltaic (PV) solar cells as one of the promising renewable energy sources which has the potential to meet the future energy demand. The importance of developing different types of renewable energy sources which include solar, wind, hydro, biomass, geothermal and hydrogen gas to supply the energy for sustainable development will present. The different types and the principle of the PV cells fully discussed. The potential applications and futures prospect of the PV solar energy system in the various area of life will be considered and discussed.

Keywords: renewable, energy, photovoltaic, solar energy, solar cells

1. Introduction

In recent years, the world has experienced a number of health pandemics [1]. A pandemic, according to the World Health Organization (WHO), is the spread of a new illness that negatively affects a sizable section of the global population [2]. However, the most recent pandemic known as the COVID-19 has given grave consequences to the whole world. Due to the dissemination of COVID-19, a variety of events that had an impact on how energy was used were seen from the perspective of the energy industry [3, 4]. Due to the release of greenhouse gases, fossil fuels including coal, oil, and natural gas have been shown to have a negative impact on both the environment and human health (GHGs). Therefore, as long as the globe continues to rely on conventional fuel-powered economy, the ambition of achieving a cleaner environment will not be possible. The world will eventually look for alternate energy since present fuels are not sustainable. In order to revitalize the economy, create more jobs, and enable the countries that are using them to be energy independent, this situation could be resolved by the efficient and effective use of renewable energy sources, which have a prodigious potential to provide a great amount of energy that exceeds the global energy demand. The share of renewable energy in the world's power generation was 28% in 2020, and it is anticipated to increase to above 26% by 2020 [5]. **Figure 1** illustrates how renewable energy is produced from resources that are renewed naturally. It is a clean and sustainable energy from natural sources such as Sun, Wind, Water, Ocean and Earth natural activities [6]. Concerns about air pollution, public

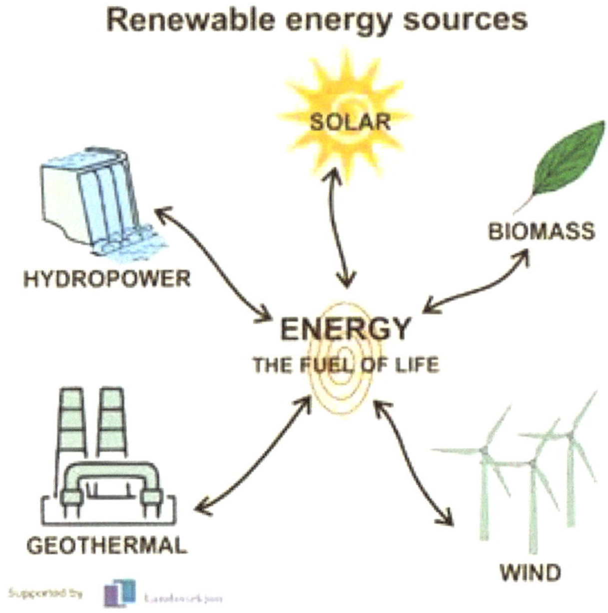

Figure 1.
Renewable energy sources.

pressure, and the desire to produce clean, livable, and egalitarian communities, among other things, all contributed to the sustained growth of government participation in 2020. In certain cases, the COVID-19 pandemic-induced global health and economic crisis has strengthened these activities, which have used a variety of objectives, policies, and actions to demonstrate their commitment to renewable energy: Around 25% of the urban population, or more than 1 billion people, resided in cities in 2020 that had either a renewable energy target or policy [7]. Since the 1990s, renewable energy technologies (RETs) have demonstrated the quickest growth rate among the various energy sources, maybe in spite of this tiny proportion (**Figures 2–5**) [9]. The focus of this chapter is to provide an adequate understanding about all the renewables, going through their share in world energy consumption, their importance, various technologies adapted to harness these renewable sources. Also an evaluation regarding the environmental impact, economics, and other social effects resulting from the use of the renewable energy systems is provided (**Table 1**).

1.1 Geothermal energy production

Geothermal energy is the residual heat developed from the formation of the earth billions of years ago as shown in **Figure 6**. The radioactive decay of potassium's radioactive isotopes causes phenomena known as radiogenic heat, which produces heat at a rate of around 3.5*109 W/kg of the element. Thorium produces 26.3*106 W/kg whereas uranium produces 96.7*106 W/kg. About half of the heat that is transferred from Earth's core to its surface is caused by this heat. Though there are pros and cons of geothermal energy (**Table 2**) [10]. There are various approaches employed by numerous people and institutions to classify the

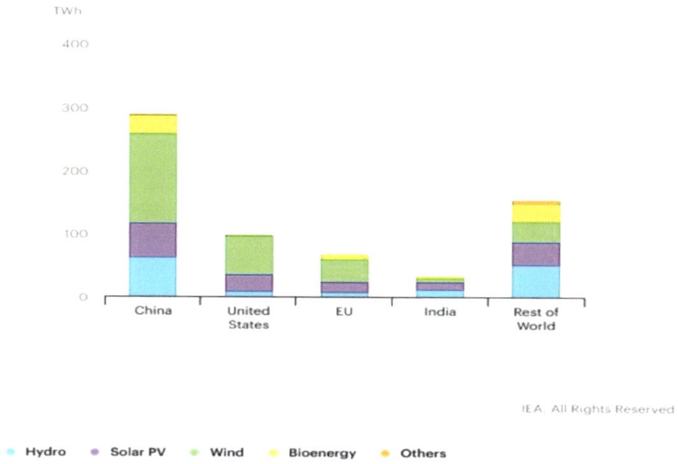

Figure 2.
Increase in renewable electricity production by technology, country, and region 2020–2021 [8].

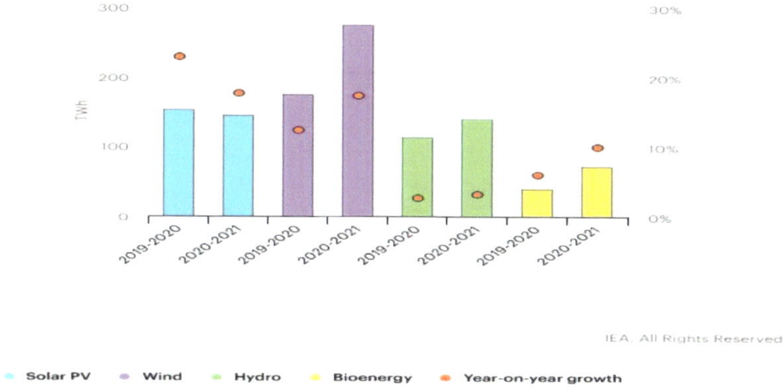

Figure 3.
Renewable electricity generation increase by technology, 2019–2020 and 2020–2021 [8].

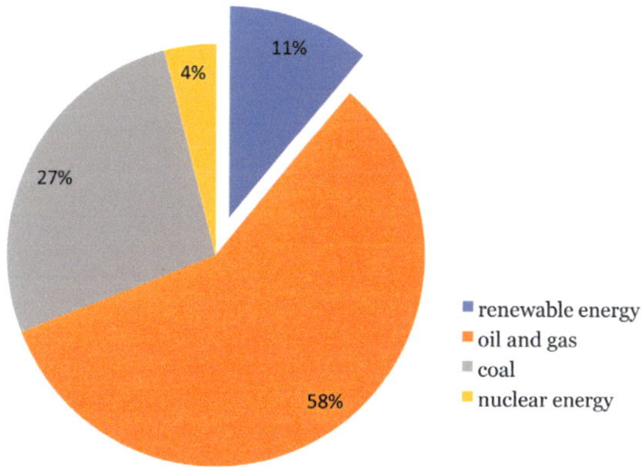

Figure 4.
World energy consumption for the year (2018).

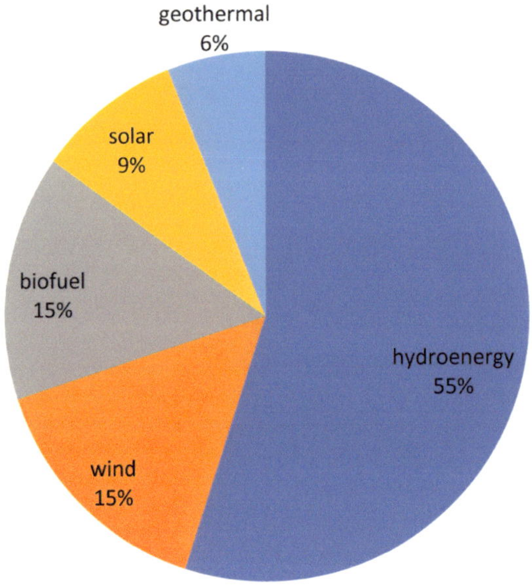

Figure 5.
Renewable energy consumption (%).

Source	Technology	Energy product
Bio-energy	domestic Combustion industrial Combustion Gasification (power and fuel production)	Heat (cooking, space heating) Process heat, steam, electricity Electricity, heat. Hydrocarbons H2
Wind energy	Sails	Movement
	Water pumping (turbines and mills)	power
	Onshore wind turbines	Electricity
	Offshore wind turbines	Electricity
Solar energy	Photovoltaic solar energy	Electricity
	Solar thermal electricity conversion	Heat, steam
	Low-temperature solar energy use	Electricity, Heat (water and space heating, cooking, doing) and cold
	Passive solar energy use Artificial photosynthesis	Heat, cold, light, ventilation H2 or hydrogen rich fuels
Geothermal energy	Geothermal conversion	Heat, steam, electricity
Hydro energy	Hydropower	Power, electricity
	Tidal energy conversion	Electricity
	Wave energy conversion	Electricity
	Current energy conversion	Electricity
Ocean thermal energy conversion	Heat, electricity	
	Salinity gradient / osmotic energy	Electricity

Table 1.
Classification of renewable production [10].

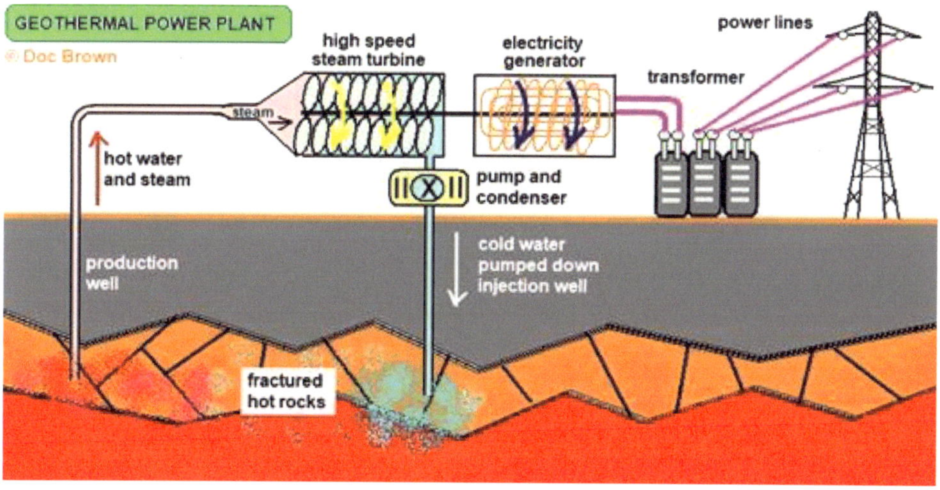

Figure 6.
Schematic diagram of geothermal power plant [11].

Pros	Cons
This energy source is more environmentally friendly than conventional fuel sources.	The largest single disadvantage of geothermal energy is that it is location specific.
A source of renewable energy.	Gases are released into the atmosphere during digging.
The number of exploitable geothermal resources will increase with on-going research and development in the industry.	Geothermal energy runs the risk of triggering earthquakes.
A sustainable source of energy as its. Always available unlike wind and solar.	Expensive resource to tap into, with high upfront costs ranging from around $2–7 million for a plant with a 1 megawatt capacity.
A reliable source as its. Easier to predict the power output from a geothermal plant with a high degree of accuracy. No fuel is required.	Energy fluid needs to be pumped back into the underground reservoirs faster than it is depleted. Management is required to maintain sustainability.

Increase in exploration meaning that new technologies are being created to improve the energy process.

Table 2.
An overview of geothermal Pros and Cons [12].

geothermal resources, the classifications based on thermal and compositional features of the resources and geothermal fluid temperature/enthalpy are the most common [13].

1.1.1 Geothermal energy application

Geothermal energy is applicable in several energy production applications. For example:

i. Geothermal energy can be exploited to provide space cooling and refrigeration via the use of absorption cooling systems.

ii. The other methods of energy production include flash steam plants, dry steam plants, binary cycle plants, and hybrid power plants. It is also employed in district heating systems and for heat generation using heat pumps.

iii. Geothermal energy may be used to produce fuel by supplying the necessary energy to run a water electrolyzer to produce hydrogen and in the manufacture of ethanol.

1.2 Wind energy

The conversion of wind energy into more usable forms, often electricity, is known as wind power. In 2005, wind power facilities with a capacity of 58,982 megawatts provided less than 1% of the world's electricity. The production of wind energy increased by more than fourfold globally between 1999 and 2005. The majority of today's wind energy is produced as electricity by using an electrical generator to transform the movement of turbine blades into an electrical current. Wind energy is employed in windmills, a much older technology, to drive mechanical equipment that performs labor-intensive tasks like pumping water or crushing grain (**Figures 7** and **8**). Both large-scale wind farms and small-scale individual turbines can generate electricity for usage in remote regions using wind power. If wind energy is utilized to

Figure 7.
Wind energy generating system [14].

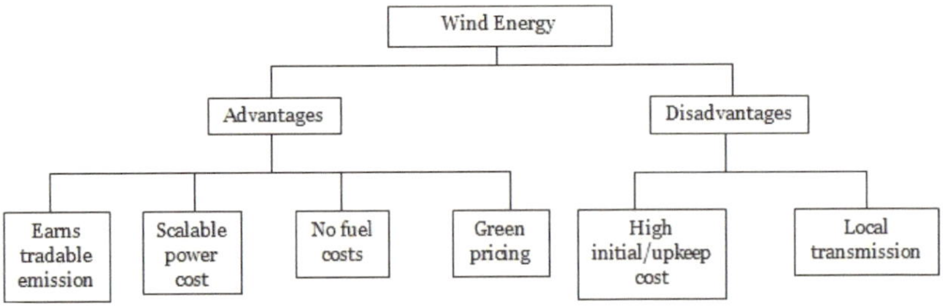

Figure 8.
Wind energy merits and demerits [15–18].

replace power produced from fossil fuels, it is abundant, renewable, widely available, clean, and reduces the greenhouse effect [14].

1.3 Hydro-energy

Hydro-energy is sourced from water activities. Water activities can be the water movement, tidal activities, ocean current and waves, physical and chemical reactions taking place in and on the water body. The water can be from surface water, ground water or an ocean. Hydro-energy can be in form of hydropower, tidal energy, wave energy, ocean salinity gradient exploitation and ocean thermal energy conversion (**Figure 9**). This hydropower is derived by using flowing water to drive turbines such that the output efficiency depends on the volume of water and the kinetic energy of the running water [20]. Tidal energy is sourced from ocean tides' movement which results from Earth-Moon rotation and the gravitational interaction with the sun [21]. Wave energy is harvested by using technology to capture the energy of waves caused by wind passing on the ocean surface and convert it to electrical energy. The energy produced by the variations in salinity between freshwater and seawater is known as salinity gradient energy. Ocean thermal energy conversion uses the temperature differential between deep cold water and warm surface water to produce power.

1.4 Bio-energy

Bio-energy is derived from the use of biomass. Biomass can be converted into different forms of bioenergy (**Figure 10**). Biomass is an organic, renewable source; it includes the entire successive species on the food chain, and also all biological waste [22].

1.5 Solar energy

Solar energy is simply energy from the sun; it is more or less the source of other forms of renewable energy, it involves capturing and harnessing the sun's energy and converting it into useful forms of energy. Solar energy is the most abundant source of electricity which in overtime its significant advancements would have

Figure 9.
Hydro energy generating system [19].

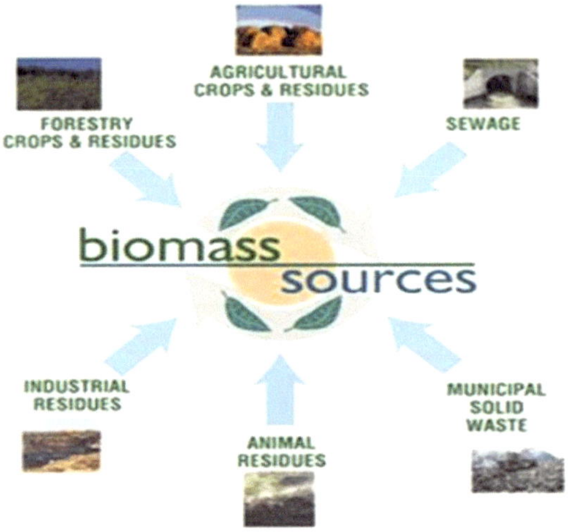

Figure 10.
Biomass energy system [22].

been produced [23–25]. There are different ways of harnessing solar energy like direct solar heating, solar radiation concentration, and solar cells.

1.5.1 PV solar panels

The photovoltaic effect is the basis for photovoltaic, which is the most direct method of converting solar energy into electricity (**Figure 11**). The appearance of an electric voltage between two electrodes connected to a solid or liquid system following the application of light to this system is known as the photovoltaic effect. Photovoltaic

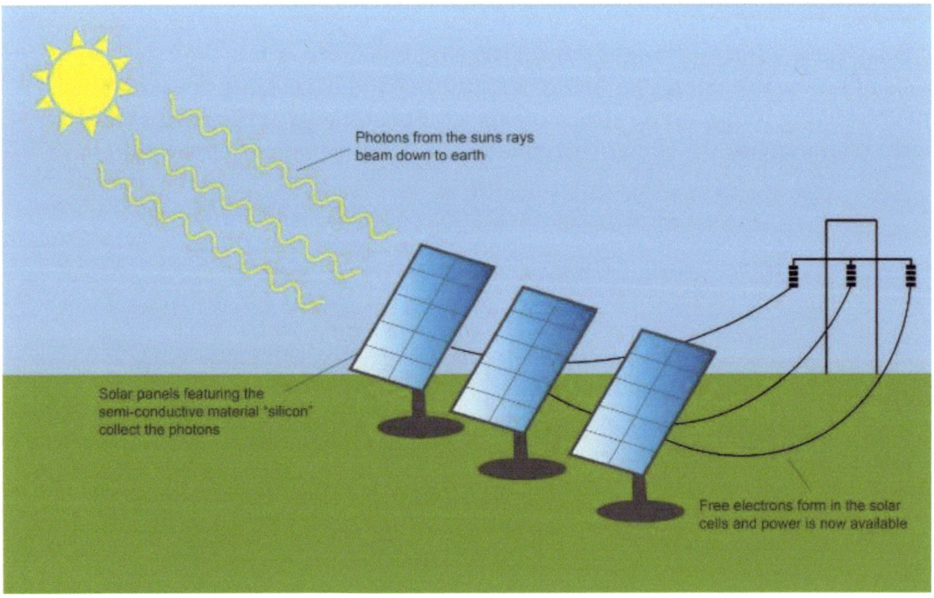

Figure 11.
Solar Electricity System [26].

of different types gratify diverse needs and tenacities. Assumed that sunlight can be used contrarily whether on Earth or in space points to the fact that location, itself, is a significant factor when it comes to selecting one of the types of photovoltaic over another. Moreover, classification by generation focuses on the materials and efficiency of the different types of PV [27, 28].

1.6 Different types of solar cells

Figure 12(**i-xxi**) [29, 30].

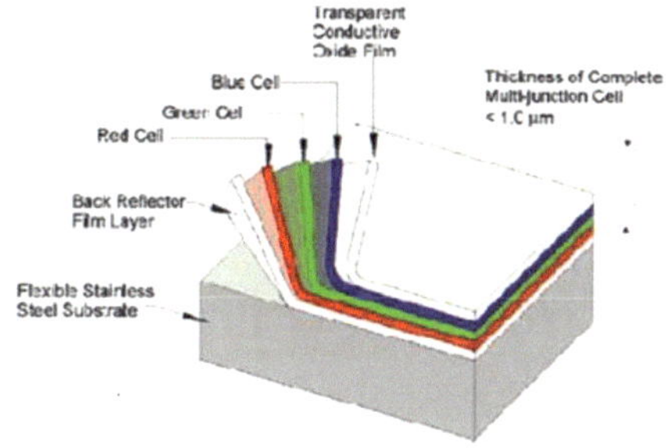

(i)

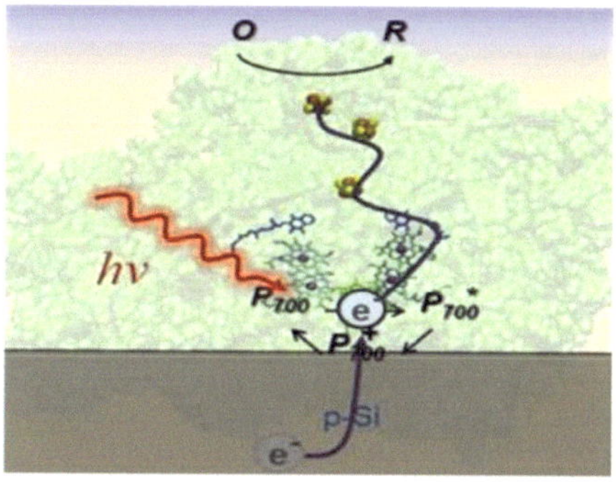

(ii)

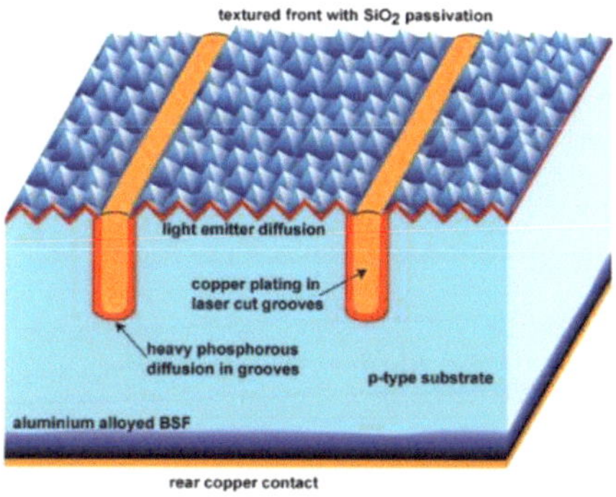

(iii)

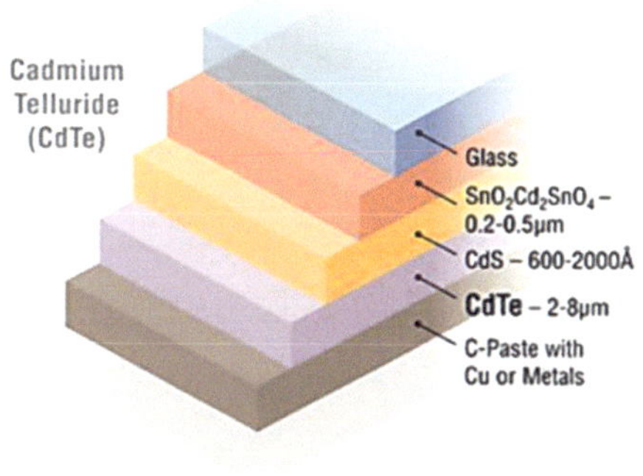

(iv)

(v)

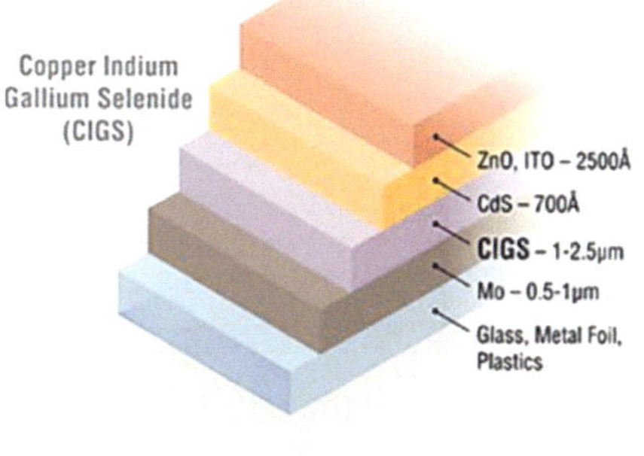

(vi)

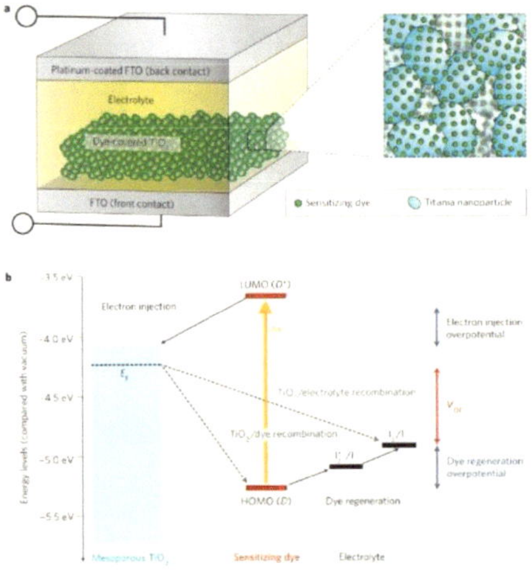

(vii)

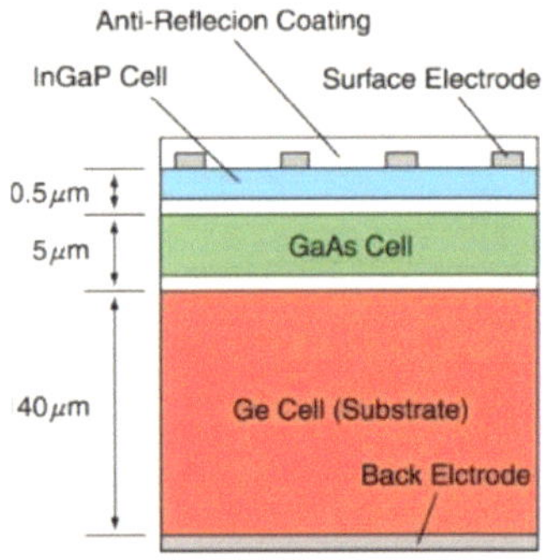

(viii)

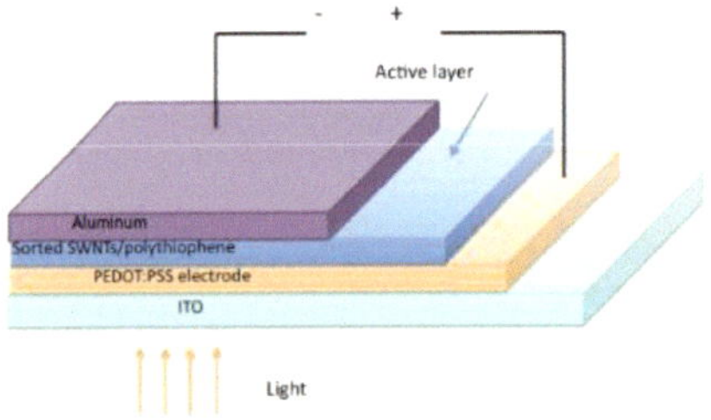

(ix)

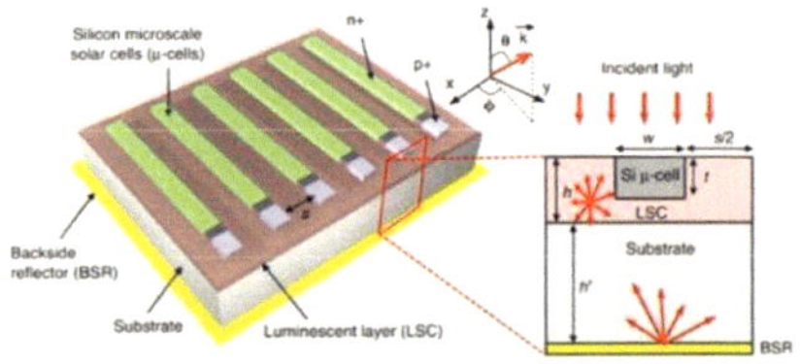

(x)

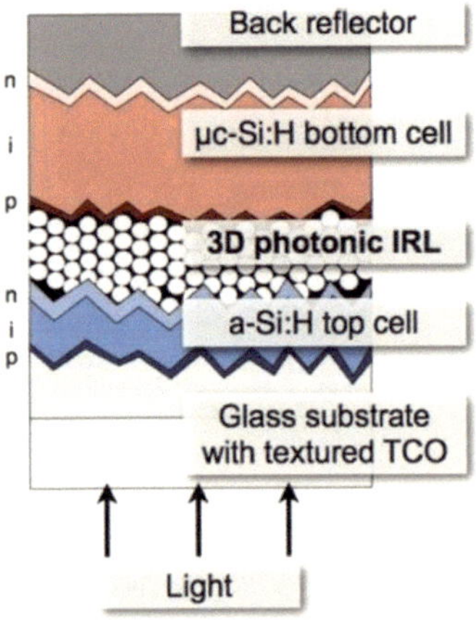

(xi)

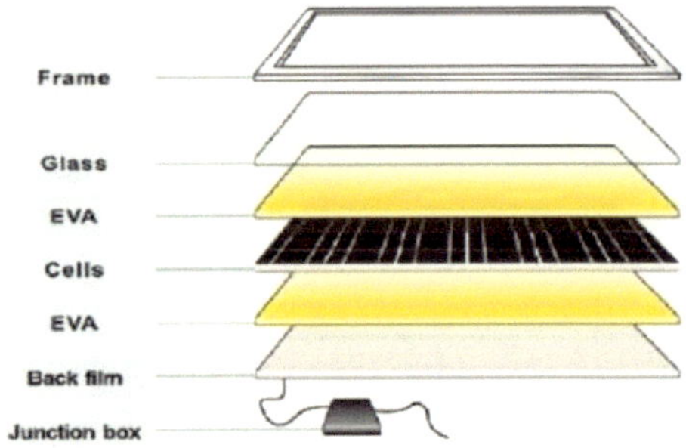

(xii)

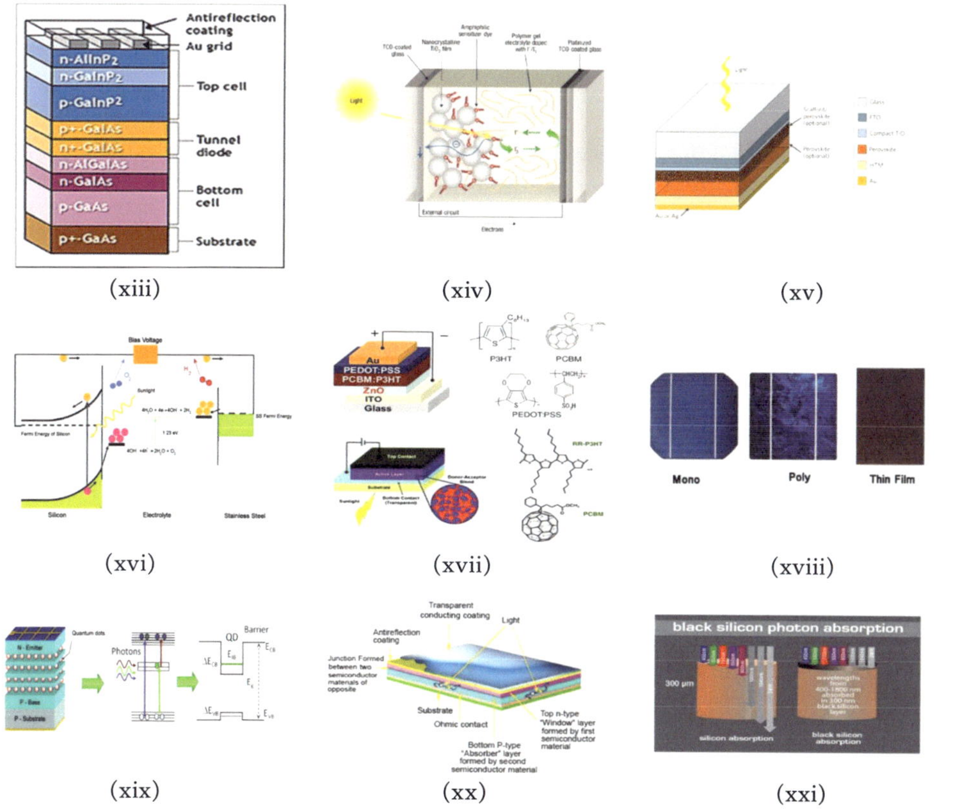

Figure 12.
Types of cells (Figures i to xxi). (i) Amorphous Silicon Solar Cell (A-Si); (ii) Biohybrid Solar Cell; (iii) Buried Contact Solar Cell; (iv) Cadmium Telluride Solar Cell (CdTe); (v) Concentrated PV cell (CVP and HCVP); (vi) Copper Indium Gallium Selenide Solar Cells (CI (G) S); (vii) Dye-Sensitized Solar Cell (DSSC); (viii) Gallium Arsenide Germanium Solar Cell (GaAs); (ix) Cell Hybrid Solar; (x) Luminescent Solar Concentrator Cell (LSC); (xi) Micromorph Cells (Tandem-Cell Using a-Si/μc-Si); (xii) Monocrystalline Solar Cell (Mono-Si); (xiii) Multijunction Solar Cell (MJ); (xiv) Nanocrystal Solar Cell; (xv) Perovskite Solar Cell; (xvi) Photoelectrochemical Cell (PEC); (xvii) Polymer Solar Cell (xviii) Polycrystalline Solar Cell (Multi-Si); (xix) Quantum Dot Solar Cell; (xx) Thin Film Solar Cell (TFSC); (xxi) Black Silicon Solar Cells.

1.7 Principle of PV solar cells

When light falls on a photovoltaic (PV) cell -also called solar cell device such light may be reflected, absorbed, or pass right through the cell.

2. Some applications of solar energy

1. Solar Power plants:

   Solar power plants: The sun's heat may boil water to produce steam that can be used to turn turbines. Solar power plants: The sun's heat may boil water to produce steam that can be used to turn turbines. To convert sunlight into electricity solar panels, photoelectric technologies and thermoelectric technologies among other can be used.

2. Homes:

   Residential appliances can easily use electricity generated through solar power. Such as solar heating system and solar drying system

3. Commercial use:

   PV modules or any other kind of solar panel can be mounted on the roofs of different buildings so as to generate electricity.

4. Ventilation system:

   Solar energy is used for ventilation purposes at many places. It is beneficial to operate bathroom, floor, and ceiling fans in buildings to reduce moisture and odor, as well as in homes to remove heat from the kitchen.

5. Power pump:

   Solar power did not just help in improving ventilation system at various homes but can also help in circulating water in any building.

6. Swimming pools:

   In any season, swimming pools are a lot of fun for both adults and children. However, keeping the water heated in these pools throughout the cold months requires a lot of energy which many people can benefit from solar energy in this regard.

7. Solar Lighting:

   These lights are also known as day lighting, and work with help of solar power. These lights capture solar energy throughout the day and turn it into electricity to illuminate at night.

8. Solar Cars:

   It is an electrical vehicle which is recharged form solar energy or sunlight.

9. Remote applications:

   Remote structures are extensively utilizing solar energy for facilities like clinics, community centers, and schools.

3. The future of PV technology

The solar energy industry is starting to move accelerative quite quickly now. With more people opting for greener ways to power their homes, the market and the consequent solar photovoltaic investigation and improvement is increasing exponentially. The researchers are doing finding on cheaper and more eco-friendly solar cells.

4. Summary and conclusion

Solar energy is the sum of the heat and light that the sun produces. This energy moves from sun and reaches the earth where human tap and collects it through solar collectors and been converted into any desirable form of energy.

Author details

Sanusi Yekinni[1,2], Ibrahim Asiata[3], Oyeshola Hakeem[1]* and Lawal Mubarak[1]

1 Department of Pure and Applied Physics, Ladoke Akintola University of Technology Ogbomoso, Nigeria

2 Nanotechnology Research Group (NANO$^+$), Ladoke Akintola University of Technology Ogbomoso, Nigeria

3 Department of Pure and Applied Chemistry, Ladoke Akintola University of Technology Ogbomoso, Nigeria

*Address all correspondence to: hooyeshola@lautech.edu.ng

References

[1] Arslan H, Bilal BMF. Contemporary research on spillover effects of COVID-19 in stock markets. A systematic and bibliometric review. Sci Forum. 2021:1-14. DOI: 10.3390/ECERPH-3-09103

[2] Badr HS, Du H, Marshall M, Dong E, Squire MM, Gardner LM. Association between mobility patterns and COVID-19 transmission in the USA: A mathematical modelling study. The Lancet Infectious Diseases. 2020;**20**(11): 1247-1254

[3] Akrofi MM, Antwi SH. COVID-19 energy sector responses in Africa: A review of preliminary government interventions. Energy Research and Social Science. 2020;**68**:101681

[4] Bashir MF, Ma B, Shahbaz M, Jiao Z. The nexus between environmental tax and carbon emissions with the roles of environmental technology and financial development. PLoS One. 2020e;**15**(11): e0242412. DOI: 10.1371/journal.pone.0242412

[5] International Energy Agency (IEA). Key World Energy Statistics 2016. 2016. DOI: 10.1787/key_energ_stat-2016-en

[6] Adeniran PO, Adejumobi CA, Awodugba AO, Sanusi YK, Oladejo DA. Overview of renewable energy technology in Nigeria power sector. International Journal of Physical Science. 2009;**4**(1):77-83

[7] Bashir MF, Sadiq M, Talbi B, Shahzad L, Bashir MA. An outlook on the development of renewable energy, policy measures to reshape the current energy mix, and how to achieve sustainable economic growth in the post COVID-19 era. Environmental Science and Pollution Research. 2022. DOI: 10.1007/s11356-022-20010-w

[8] Available from: iea.org/reports/global-energy-review-2021/renewables

[9] German MP. Renewable energy sources (basics) a petro-state using renewable energies. 2008. DOI: 10.1007/978-3-531-91003-1-4

[10] Turkenburg WC, et al. World energy assessment: Energy and the challenge of sustainability. 2000. ISBN: 92-1-126126-0

[11] Available from: www.docbrown.info/ephysics/energy3.htm

[12] Available from: https://www.twi-global.com/

[13] Bronicki LY. Geothermal power conversion technology. In: Meyers RA, editor. Encyclopedia of Sustainability Science and Technology. New York, NY: Springer; 2012. pp. 4234-4339

[14] Available from: https://www.energy.gov/eere/wind/advantages-and-challenges-wind-energy

[15] Amusan JA, Sanusi YK, Fajinmi GR. Determination of annual energy captures potential for wind power system. Research Journal of Applied Science. 2007;**2**(9):927-930

[16] Sanusi YK, Abisoye SG. Estimation of wind energy potential in Southwestern Nigeria. The Pacific Journal of Science and Technology. 2011; 12(2):160-166. Available from: http://www.akamaiuniversity.us/PJST.htm

[17] Sanusi YK, Adedokun O. Determination of energy production potential of wind resources in

LAUTECH Ogbomoso, Nigeria. Journal of Engineering and Applied Sciences. 2012;4:34-39. Available from: www.cenresin.org.publications

[18] Latunji S, Sanusi YK. Appraisal of wind energy potential in Zaria metropolis. Global Journal of Science Frontier Research. 2013;13(1):19-27. Thomson Reuters Impact Factor: 2.42 USA. www.journalofscience.org/index.php/GJSFR

[19] Available from: https://www.conserve-energy-future.com/hydroelectricpower.php

[20] International renewable energy agency (IRENA). irena.org retrieved May 17, 2022

[21] Turcotte DL, Schubert G. Chapter 4. In: Geodynamics. 2nd ed. England: Cambridge; 2002

[22] Available from: https://www.google.com/url?sa=i&url=https%3A%2F%2Fwww.bioenergyconsult.com%2Fa-glance-at-biomass-energy%2F&psig=AOvVaw2zMkNDlS6j01xP7Pclxhjt&ust=1673770833860000&source=images&cd=vfe&ved=0CA0QjRxqFwoTCMCa3rPQxvwCFQAAAAAdAAAAABAW

[23] Dutta D, Habeeb O, Usman AJA. Sustainable energy in rural communities of Bongouanou: Utilizing solar energy as a source for electricity. In: IEEE Global Humanitarian Technology Conference: South Asia Satellite (GHTC-SAS), Trivandrum. 2013. pp. 15-20

[24] Balogun SW, James OO, Sanusi YK, Oyeshola HO. Green synthesis and characterization of zinc oxide nanoparticles using bashful (Mimosa pudica), leaf extract: A precursor for organic electronics applications. SN Applied Sciences. 2020;**2**:504. DOI: 10.1007/s42452-020-2127-3

[25] Ogundeji ST, Awodele MK, Oyeshola HO, Adedokun O. Optical studies of titanium dioxide/silver/gold (TiO_2/Ag/Au) nanocomposites as photo anode in dye sensitized solar cells. IOP Conference Series: Materials Science and Engineering. 2020;**805**:012027. DOI: 10.1088/1757-899X/805/1/012027

[26] Available from: https://www.clean-energy-ideas.com/solar/solar-panels/solar-panel-diagram/

[27] Oyeshola HO, Adisa MA, Adejumo BK, Babalola KK, Agboluaje BA, Adedokun O, et al. Effect of low temperature synthesis of carbon nanotube nanocomposite on the photovoltaic performance of anode buffer layer in polymer solar cell. IOP Conference Series Materials Science and Engineering. 2020;**805**:012026. DOI: 10.1088/1757-899X/805/1/012026

[28] Yoshikawa K, Kawasaki H, Yoshida W, Irie T, Konishi K, Nakano K, et al. Silicon heterojunction solar cell with interdigitated back contacts for a photoconversion efficiency over 26%. Nature Energy. 2017;**2**:17032

[29] Luque A, Hegedus S. Handbook of Photovoltaic Science and Engineering. Chichester, West Sussex, United Kingdom: John Wiley & Sons Ltd, Wiley; 2011

[30] Bagher AM, Vahid MMA, Mohsen M. Types of solar cells and application. American Journal of Optics and Photonics. 2015. DOI: 10.11648/j.ajop.20150305.17